A CENTURY OF ASTRONOMY IN THE JOURNAL OF THE WASHINGTON ACADEMY OF SCIENCES

ISBN

978-0-615-61836-4

Preface

The Washington Academy of Sciences incorporated in 1898 as an affiliation of eight Washington D.C. area scientific societies. The founders included Alexander Graham Bell and Samuel Langley, Secretary of the Smithsonian Institution. The purpose of the new Academy was to encourage the advancement of science and "to conduct, endow, or assist investigation in any department of science." That purpose which guided the Academy throughout its first 100 years will continue to be our guide through the coming century.

As we move through our second century we now publish scientific monographs which consist of reprints from our peer-reviewed publications – the *Proceedings* (from 1899 to 1910) and the *Journal* (1911 to the present), each monograph dedicated to a specific field of science.

This first monograph we dedicate to the field of astronomy, a field in which the Academy has over 100 years of coverage. From 1905 to 2011 it is our pleasure to present a wide variety of astronomical papers, eleven in all. Each paper is preceded by a short description of the science described in that paper. Some of the figures and photographs have faded through the years, and so our scans cannot always reproduce the original quality. We hope you enjoy this glimpse of past astronomy at the Academy.

August 2012

Sethanne Howard, editor

Table of Contents

Introduction to

Declinations of Certain North Polar Stars Determined with the Meridian Circle

Harriet Bigelow (1870 – 1934) was an instructor in astronomy at Smith College. She served as a President of the AAVSO (American Association of Variable Star Observers). The following article appeared in 1905 in the Proceedings of the Washington Academy of Sciences. Omitted are the sixty pages of tables that accompany the article. If these tables are desired please contact the Washington Academy of Sciences for a scan of the original paper.

We include the article here to illustrate the intricate nature of measuring stellar coordinates. Declination is celestial latitude (0° to ±90°). Right ascension is celestial longitude (0^{hrs} to 24^{hrs}). Near the celestial pole (declination = ±90°) measurements are difficult because the differences in right ascension angles are so small. Hence the need for a plethora of stars measured near the pole. Although the techniques described refer to now obsolete equipment, equally intricate measurements and data reduction methods are used for modern observations such as those taken with the astrometry satellite Hipparcos (operated 1989 – 1993) which measured stellar positions for over a million stars.

Declinations of Certain North Polar Stars Determined With the Meridian Circle

Harriet W. Bigelow

Smith College

The stars whose right ascensions and declinations I have observed with the Meridian Circle are those requested by Dr. Auwers in the *Astromische Nachrichten*, No. 3440. They comprise a list of 21 stars between 84° 34′ and 88° 55′ north declination and in magnitude ranging from 5.3 to 7.5. As Dr. Auwers points out, the *Berliner Jahrbuch* at present gives but 10 stars of declination above 82°, 5 of these being above 85°; and these are not symmetrically placed in right ascension leaving several gaps of 2 or 3 hours when an observer would find no fundamental star of high declination available. The present observations were undertaken to furnish accurate places of additional stars of high declination for use when such are needed in determining instrumental constants.

The observations were begun in October, 1901, and extended through the period to the end of June, 1903. The right ascensions have not yet been reduced.

The Walker Meridian Circle was built by Pistor and Martins of Berlin in 1854. The telescope tube is heavy, unsymmetrical, and shows considerable flexure; the object-glass and eye-ends are not interchangeable, as in many modern instruments. The objective, of 6.3 inches aperture, was examined at the Physical Laboratory. The focal length, 251.6 cm. or 8 ft. 0.8 in., was determined by measurements on the negative unit planes. The radii of curvature, measured with the spherometer, were found to be 165.7 cm. for the outer curve, 274.4 cm. for the inner curve. The structure of the glass was examined by means of Nicol prisms at conjugate foci. For perpendicular position of the prisms the lens instead of being entirely dark shows irregular light portions extending toward the center, due to irregular polarization in the glass. Practically, however, the lens gives excellent star images for meridian circle work, *i.e.*, small, round disks, of uniform size across the entire field.

The graduated circles of the instrument are 37 1/2 inches in diameter. The fine circle, which was the one employed, is graduated to 2′ and is read by 4 microscopes of 16 magnifying power reading to tenths of a second of arc. Each microscope has been furnished with two sets of threads one and a half revolutions apart to eliminate periodic error. For a reading two divisions of the circle were pointed on, the micrometer screw being turned always one half revolution. The readings were corrected for error in the run.

The micrometer eye-piece was obtained a few years ago from the Repsolds. It contains 25 vertical threads in groups of 5, and 2 horizontal threads about 5″ apart. There is no declination micrometer screw. Settings were made with the tangent screw of the instrument, bringing the star to the point half-way between the horizontal threads. It was usually found possible to make 3 or more pointings with the corresponding readings of the microscopes while the star was crossing the field. The positions off the meridian were symmetrically chosen to avoid error caused by possible inclination of the wires. The reductions to the meridian were made according to the formula

$$z = z' - \sin 2\delta' \frac{\sin^2 \frac{1}{2}t}{\sin 1''}$$

where δ' is the apparent declination. In this form a second term becomes negligible. (See Leyden Observations, Vol. VI, p. LX.) Tables were made out for each star from which the correction could be taken with the declination and hour angle as arguments.

Observations for nadir were made about every 3 hours. These were obtained by turning the telescope over a mercury basin and observing the reflection of the horizontal threads by means of a collimating eye-piece. Four settings were made, the mercury basin being turned 180° in the middle of the set. When successive nadirs differed by more than 0″.50, it was assumed that the difference was directly proportional to the time; when the difference was less than 0″.50 the straight mean was taken. Occasionally during observations for nadir the instrument seemed to move after a setting had been made, showing either that it was under a strain, or possibly that the surface of the mercury changed slightly.

The plan was to obtain for each star, both at upper and lower culminations, 2 observations in each of the four following positions: clamp west, direct; clamp west, reflected; clamp east, direct; clamp east, reflected. This plan was not entirely carried out as the tables show, in part due to the difficulty in obtaining reflected observations. These were often prevented by wind or unsteady seeing. Often, too, reflected observations were prevented by trains on the Michigan Central Railroad, and sometimes by the shutting of a door in another part of the building. Nevertheless, nearly as many reflected observations were obtained as direct. They seem to be quite as consistent among themselves as the direct; perhaps they are in a sense selected observations. Each night's observing list included at least one of the *Berliner Jahrbuch* stars.

Advantage in combining reflected and direct observations is found in the fact that different sets of divisions on the circle are employed, thus largely eliminating division errors, and in the fact that the sign of the sine flexure is reversed. In the mean of the 4 positions : W. D. ; W. R. ; E. D. ; E. R. ; the first 2 terms of the cosine flexure and the first term of the sine flexure are eliminated.

To determine the amount of the flexure the following formulae were employed:

W.D. $\qquad \zeta = z_1 + a' \cos z + b' \sin z - (180° + N) + a'$

W.R. $(180° - \zeta) = z_2 - a' \cos z + b' \sin z - (180° + N) + a'$

E.D. $(360° - \zeta) = z_3 + a' \cos z - b' \sin z - (180° + N) + a'$

E.R. $(180° + \zeta) = z_4 - a' \cos z - b' \sin z - (180° + N) + a'$

The coefficient of cosine flexure found was $1''.694$; and of sine flexure, $0''.117$. In the case of clamp west the circle readings increase from the zenith toward the north and the formula for flexure correction is

$$\zeta = z + 0''.162 - 1''.694 \cos z + 0''.1117 \sin z - 1''.694$$

The large cosine flexure was also found by Dr. Hall. (See "Reprint from Report of Michigan Academy of Science," 1904.)

Corrections for division errors were not applied. Some examination of the circle was made to determine its general character. The mean of the 4 divisions, 0°, 90°, 180°, 270°, was assumed to be without error. In finding the errors of the intermediate divisions, taking also as a division the mean of 4 marks 90° apart, 2 microscope arms were set 100° apart, 110° apart, etc. Readings were then made on a number of 100° spaces, for instance, distributed symmetrically around the circle; these readings were taken forward and back so as to eliminate progressive changes in the instrument depending on the time. The mean of these readings was assumed to be the correct 100° space and was used for obtaining the error of the 100° mark on the circle. The microscopes were afterward changed 180° from their first position and the process repeated.

4

The 2 columns of division errors given show the changes produced by placing the microscopes in the two positions. Evidently the effect of gravity is considerable as might be expected from the structure of the circle, which is rather frail.

Divisions		Division Errors		Means
100	(vs. 10° etc.)	−0″.11	−0″.62	−0″.36
110	20	−0.33	−0.84	−0.58
120	30	−0.50	−0.58	−0.54
130	40	−0.72	−0.62	−0.67
140	50	−0.70	−0.62	−0.66
150	60	−0.34	−0.94	−0.64
160	70	+0.06	−0.26	−0.10
170	80	+0.40	+0.02	+0.21
180	90	0.00	0.00	0.00

Two tables of observed declinations are presented, the first giving the absolute declinations from the circle readings without correction for flexure or division errors; the second giving the declinations of the list stars from comparison with the one or more zero stars observed on the same night.

In Table I the observed zenith distances are given, corrected for runs, reduction to the meridian, and refraction. Bessel's refraction tables were employed, as prepared by Professor Eastman of the Naval Observatory. The standard barometer was repaired and tested a few years ago by the Weather Bureau Office in Washington. The thermometers also have been tested by them and by the weather bureau official at Lansing. During observations the thermometer was hung near the object glass of the telescope, and the readings were corrected when necessary, according to the table of corrections determined by the Weather Bureau Office.

The next column in Table I gives the observed zenith distances, reduced to Jan. 0.0 of the year of observation. These reductions were made with the "Independent Star Numbers" G, H, etc., given for each day in the *Berliner Jahrbuch*. They were checked by a sufficient number of identical reductions made with the Besselian Star Numbers, A, B, C, D, E, from the *Berliner Jahrbuch* and the star constants from Dr. Auwers's list in the *Astronomische Nachrichten*.

The zenith distances are then reduced to Jan. 0.0, 1900 and the means taken of the different observations of each position. In obtaining these means, a system of weights depending on the number of settings in each case was adopted as follows: probable error of one setting $0''.33$; probable error of nadir determination $0''.25$; probable error in refraction tables $0''.30$, giving as weights:

No. of settings	Weight
1	.72
2	.90
3	1.00
4	1.04
5	1.08
6	1.10
7	1.11

The reflected observations are also corrected for the position of the mercury basin, the correction being $h \tan z$ where h is height of telescope axes above the artificial horizon. This correction is $0''.04$ for upper culmination, and $0''.03$ for lower.

The mean of the 4 positions above pole combined with the mean of the 4 below gives the value for latitude corresponding to each star.

LATITUDE.

List Stars.		B. J. Stars.	
Cephei Br. 256	49″.35	43 H. Cephei	48″.74
Cephei 157 Hs.	48.95	Polaris	48.82
Cephei 158 Hs.	48.86	Gr. 750	48.68
Cephei 109 Hs.	48.72	51 H. Cephei	48.80
Urs. min. 4 B	48.71	1 H Draconis	48.76
Cephei 121 Hs.	48.48	30 H. Camelop.	49.10
Urs. min. 3 Hs.	48.42	δ Urs. min.	48.94
32 H. Camel. pr.	48.70	λ Urs. min.	48.65
" " " seq.	49.08	76 Draconis	48.80
Cephei 135 Hs.	48.75	Mean	48.81
Urs. min. 57 B	48.74		
Cephei 3 Hs.	48.60		
Cephei Gr. 3548	48.60		
32 H. Cephei	48.72		
36 H. Cephei	48.58		
39 H. Cephei	48.61		
Cephei 125 Hs.	48.72		
Mean	48.74		

The nine zero stars give 48″.81; the seventeen others, 48″.74. The value from the nine was given half weight and the adopted value for the latitude of Ann Arbor is 42°16′48″.76.

The value found recently by Dr. Hall is 48″.8 (see *Astronomical Journal*, 518).

This value of the latitude combined with the zenith distances gives for each star the eight values of declination in the last column of Table I. The mean of the four values above pole, with the mean of the four below, gives the final value of absolute declination.

In the case of the five stars not observed in all eight positions, adopted values of declination were found by correcting for flexure the places obtained and combining them with arbitrary weights as follows:

Cephei 147 Hs. ¼(W. D. + 2 W. R. + E. D.) for declination above pole, combined with equal weight with position below pole.
Cephei 149 Hs. ¼(2 W. D. + W. R. + E. R.) for declination above pole, and then treated like preceding star.
Camelop. s 664. Mean of the four positions above pole.
Urs. min. 33 Hs. 1/2(W. D. + E. D.) for position below pole, combined with half weight with observations above pole.

7

s Urs. min. W. D. below pole combined with 1/3 weight with the mean of the remaining observations.

In Table II in comparing the stars of the list with the zero stars observed on the same night, differential flexure was applied. No attempt has been made to give weights to the means depending on the number of zero stars employed. The final declinations obtained by the two methods are found to agree closely.

Table III gives a summary of the observed declinations together with the declinations given in Newcomb's "Fundamental Catalogue of Stars," and those given in the *Berliner Jahrbuch* for 1900, so far as the observed stars are found in either catalogue. The *Berliner Jahrbuch* for 1906 gives also in the appendix definitive corrections to the places as given in the main catalogue. The last column of Table III has been formed by adding these corrections to the catalogue places, and reducing from 1906 to 1900, employing the *Berliner Jahrbuch* values for precession without including proper motion. In the first column of observed declinations the five values obtained, as described above, from incomplete sets of observations are bracketed.

UNIVERSITY OF MICHIGAN, ANN ARBOR, May, 1904.

Introduction to

On "earth light," or the Brightness Exclusive of Starlight, of the Midnight Sky

W. J. Humphreys was a meteorologist with the U.S. Weather Bureau in Washington, D.C. He was closely associated with three universities: Washington and Lee, from which he held three degrees, the University of Virginia where he was both a student and an instructor, and Johns Hopkins University where he received a PhD in physics. The following article appeared in 1912 in the Washington Academy of Sciences Journal, Volume 2.

This short article addresses the issue of meteoric bombardment as a possible source for the excess light in the night sky. Light in the night sky comes from several sources: direct starlight, scattered starlight, auroral light (light originating in the upper atmosphere), the polar aurora, zodiacal light, light from galactic sources, and light scattered from terrestrial sources (light pollution). Excess light is other than starlight. The "earth light" he describes in this article is not the zodiacal light which appears near sunset and sunrise and represents light scattered off dust in the ecliptic plane. Instead it probably fits within the auroral light.

On "earth light," or the Brightness, Exclusive of Starlight, of the Midnight Sky

W. J. Humphreys

U.S. Weather Bureau
To appear in the Astrophysical Journal

Investigations begun some years ago by Newcomb,[1] and continued by Yntema,[2] Fabry,[3] Abbot[4] and others, have conclusively shown that there is more light in the midnight sky than can be accounted for by the stars alone. It is not due to nebulae or any other constant source since its brightness varies from night to night and even during the same night; nor can it be caused by anything entirely outside the atmosphere, since it increases in brightness as the horizon is approached.

It has been suggested by Yntema that it may be due, at least in part, to a permanent Aurora, and indeed this is highly probable from the fact that the green "auroral line," $\lambda 5770$, may be seen on almost any dark clear night in any part of the sky.[5]

But there is another possible source of sky light, possibly of the "permanent aurora" itself, that deserves consideration; namely, the bombardment of the outer atmosphere by material of meteoric origin. So far as such a bombardment produces light at all it must be through a considerable depth of the rarer portion of the atmosphere, and therefore it should appear brighter as the zenith distance is increased.

For simplicity of numerical calculations it will be assumed that "earth light" is both constant and uniform—the same over all parts of the sky, invariable, and continuous. It will also be assumed that whatever the size of meteoric masses, (doubtless the vast majority are but minute grains) their light producing efficiency, or ratio of luminous to total energy, is the same as that of the sun.

---

[1] Astrophysical Journal, **14**: 297. 1901.

[2] On the Brightness of the Sky and the Total Amount of Starlight. Gronigen: Gebroeders Hoitsema. 1909.

[3] Astrophysical Journal, **31**: 394. 1910.

[4] Annual Report Smithsonian Institution, 1911, p. 64.

[5] Campbell, Astrophysical Journal, **2**: 162. 1895.

With these assumptions it is possible to compute, from known data, the rate at which meteoric material must be picked up to produce the observed amount of "earth light," as follows:

"Earth light" per 10 square degrees = star of first magnitude.

Full moon = star of −11.77 magnitude = 120,000 stars of first magnitude.

Area full moon = 0.2 square degree.

Hence, brightness full moon = 6 x 10^6 brightness of "earth light."

But the brightness of the full moon is equal to that of a white-mat surface illuminated by a 1200 candle-power light at one meter's distance,[6] or, in symbols, 1200 m.c. (meter candles).

Hence brightness of earth light = 2×10^{-4} meter candles.

Now normal zenith sunshine = 10^5 m.c., or is 5×10^8 times brighter than "earth light," and consequently delivers 25×10^7 times as much energy per square centimeter as would be radiated from both sides combined of a self-luminous shell equivalent in brightness to "earth light."

Hence, since the solar constant is about 1.92 calories per square centimeter per minute, the total energy used, according to the above assumptions, in the production of "earth light" is

$$4\pi R^2 \times \frac{1.92}{25 \times 10^7} \text{ calories per minute,}$$

in which R is the radius of the earth in centimeters, or

$$27 \times 10^{15} \text{ ergs per second, roughly.}$$

Let this energy be supplied by M grammes of matter moving with the average velocity of meteors, or 42 kilometers per second, then

$$\frac{1}{2} MV^2 = 27 \times 10^{15}$$

or $M = 3 \times 10^3$ roughly.

This is less than three times the amount of meteoric material Young[7] assumes as allowable, and, so far as there is any present means of

---

[6] Circular of the Bureau of Standards, **28**: 7. 1911.
[7] *General Astronomy*, p.475

knowing, may be even less than the actual amount caught up by the earth per second. Indeed it is so small that it would take about two hundred million years for it to increase the radius of the earth a single centimeter!

Numerical calculations, therefore, show that, though not proved, it is within the bounds of reason to assume "earth light" somehow due to bombardment of the outer atmosphere by fine meteoric material; and hence the possible effect of such bombardment should be taken into account in the planning of much needed further observations.

Introduction to

A Comparison of Stellar Radiometers and Radiometric Measurements on 110 Stars

William Weber Coblentz (1873 – 1962) was an American physicist notable for his contributions to infrared radiometry and spectroscopy. He headed the radiometry section at the Bureau of Standards (now the National Institute of Standards and Technology). He was among the first, if not the first, to verify Planck's Law.

The following article appeared in 1915 in the Washington Academy of Sciences Journal Volume 5. In 1913, Coblentz developed thermopile detectors and used them at Lick Observatory to measure infrared radiation from 110 stars, and the planets Mars, Venus, and Jupiter. This short article describes the results.

Today this kind of study is called absolute spectrophotometry where a standardized blackbody source is compared to a star's radiation across the energy spectrum. The telescope alternates between observing a star and a blackbody source located some distance away from the telescope – far enough so that the source appears "star-like". This allows the astronomer to calibrate the radiation from the star and put it on an absolute scale. Bolometers have much improved since 1913.

A Comparison of Stellar Radiometers and Radiometric Measurements on 110 Stars[1]

W.W. Coblentz

Bureau of Standards

In this paper experiments are described showing that there is but little difference in the radiation sensitivity of stellar thermocouples constructed of bismuth-platinum, and thermocouples of bismuth-bismuth + tin alloy, which have a 50 per cent higher thermoelectric power. Improvements in the method of maintaining a vacuum by means of metallic calcium are described whereby it will be possible to go to the remotest stations for making radiation measurements without carrying an expensive vacuum pump. With this outfit measurements were made on the radiation from 112 celestial objects, including 105 stars. This includes measurements on the bright and the dark bands of Jupiter (also a pair of his satellites), the rings of Saturn, and a planetary nebula. Quantitative measurements were made on stars down to the 5.3 magnitude; and high grade qualitative measurements were made on stars down to the 6.7 magnitude. It was found that red stars emit from 2 to 3 times as much total radiation as blue stars of the same photometric magnitude.

Measurements were made on the transmission of the radiations from stars and planets through an absorption cell of water. By this means it was shown that, of the total radiation emitted, the blue stars have about two times as much radiation as the yellow stars, and about three times as much radiation as the red stars, in the spectral region to which the eye is sensitive.

A stellar thermocouple and a bolometer were compared and the former was found to be the more sensitive. The conclusion arrived at is that, from the appearance of the data at hand, greater improvements are to be expected in stellar thermocouples than in stellar bolometers.

The object of the investigation was to obtain some estimate of the sensitivity required in order to be able to observe spectral energy curves of stars. The radiation sensitivity of the present apparatus was such that, when combined with a 3-foot reflecting telescope, a deflection of 1 mm. would have resulted when exposed to a candle

---

[1] Detailed paper to appear in the Bulletin of the Bureau of Standards.

placed at a distance of 53 miles. In order, however, to do much successful work on stellar spectral energy curves, a sensitivity one hundred times this value is desirable. In other words, assuming that the rays are not absorbed in passing through the intervening space, the radiometric equipment (radiometer and mirror) must be sufficiently sensitive to detect the radiation from a candle removed to a distance of five hundred miles. This can be accomplished by using a 7-foot mirror and by increasing the sensitivity of the present radiometer (thermo-couple and galvanometer) twenty times. This increase in sensitivity is possible.

Measurements were made to determine the amount of stellar radiation falling upon 1 cm^2 of the earth's surface. It was found that the quantity is so small that it would require the radiations from *Polaris* falling upon 1 cm^2 to be absorbed and conserved continuously for a period of one million years in order to raise the temperature of 1 gram of water 1°C. If the total radiation from all the stars falling upon 1 cm^2 were thus collected and conserved it would require from 100 to 200 years to raise the temperature of 1 gram of water 1°C. In marked contrast with this value, the solar rays can produce the same effect in about one minute.

The Distances of the Heavenly Bodies

W. A. Eichelberger was a member of the U. S. Naval Observatory staff. The following article appeared in 1916 in the Washington Academy of Sciences Journal, Volume 6.

After a brief history of measurements of the Astronomical Unit (AU, the average distance of the Earth to the Sun) the article launches into the discovery of the asteroid Eros and how it was used to determine the value of the AU – this time achieving the modern value. The article then moves to the history of stellar parallax studies showing that "It is the first positive proof that the distances of the stars are sensibly less than infinite." For millennia the lack of an observed parallax for celestial objects was used to set the celestial sphere at infinity. Today, thanks to the work of many astronomers measuring parallaxes, we know that each star has its unique (non-infinite) distance from Earth.

The phrase (Diagrams shown) that appears in the article several times refers to diagrams shown during the presentation. They are not available for the article.

The Distances of the Heavenly Bodies[1]

W. A. Eichelberger

U. S. Naval Observatory

A year ago our retiring president took the members of the Society into his confidence as follows:

> Cognizant of the fact that my election to the presidency of the Philosophical Society a year ago, obligated me to give an address of some sort one year later, I confidently waited for the inspiration that I felt would suggest a fitting subject for the occasion. The expected inspiration did not, however, materialize.

Undoubtedly because of that fact, and out of the goodness of his heart, towards the close of his address he turned to the present speaker, then presiding, and said:

> I have said nothing whatever about the determination of the distance between the planets nor of the units used by astronomers in reckoning distances of the stars. . . . They form, so to speak, other chapters of the subject which I shall leave to some future ex-president of our Society.

This call, I suppose, was intended to take the place of an inspiration, and wherever I have gone during the past twelve months the call has ever been ringing in my ears. The subject of the evening is presented therefore not as a matter of choice but from compulsion.

Before any attempt was made by the ancients to determine the distance from the Earth of any celestial body, we find them arranging these bodies in order of distance very much as we know them today, assuming that the more rapid the motion of a body among the stars the less its distance from the Earth; the stars, that were supposed to have no relative motions, were assumed to be the most distant objects.

The first attempt to assign definite relative distances to any two of the bodies was probably that of Eudoxus of Cnidus who, about 370 B.C., supposed, according to Archimedes, that the diameter of the Sun was nine times greater than that of the Moon, which is equivalent to saying that, since the Sun and the Moon have approximately the same apparent

---

[1] Presidential address before the Philosophical Society of Washington on March 4, 1916

diameter, the distance of the Sun from the Earth is nine times greater than that of the Moon.

A century later, about 275 B.C., Aristarchus of Samos gave a method of determining the relative distances of the Sun and Moon from the Earth as follows: When the Moon is at the phase first quarter or last quarter, the Earth is in the plane of the circle which separates the portion of the Moon illuminated by the Sun from the non-illuminated part, and the line from the observer to the center of the Moon is perpendicular to the line from the center of the Moon to the Sun. (Diagram shown.) If, at this instant, the angular separation of the Sun and Moon is determined, one of the acute angles of a right-angle triangle—Sun, Moon, and Earth—is known, from which can be deduced the ratio of any two of the sides, as, for instance, the ratio of the distance from the Earth to the Moon to that from the Earth to the Sun. Aristarchus gives the value of this angle as differing from a right angle by only one-thirtieth of that angle, *i.e.* it is an angle of 87°, from which it follows that the distance from the Earth to the Sun is nineteen times that from the Earth to the Moon. This method of Aristarchus is theoretically correct, but in determining the angle at the Earth as being 3° less than a right angle, he made an error of about 2° 50'.

Hipparchus, who lived about 150 B.C. and was called by Delambre the true father of astronomy, attacked the problem of the distances of the Sun and Moon through a study of eclipses. Assuming in accordance with the result of Aristarchus that the Sun is nineteen times as far from the Earth as the Moon, having determined the diameter of the Earth's shadow at the distance of the Moon and knowing the angular diameter of the Moon he found 3' as the Sun's horizontal parallax. By the Sun's parallax is meant the angle at the Sun subtended by the Earth's semi-diameter and if a = the semi-diameter of the Earth, Δ = the distance to the Sun, and II = Sun's horizontal parallax, the relation (diagram shown) between these quantities is expressed by the equation:

$$\sin \Delta = \frac{a}{II}$$

The next attempt to determine the distance of a heavenly body was made about 150 A.D. by Claudius Ptolemy, the last of the ancient astronomers and one whose writings were considered the standard in things astronomical for fifteen centuries. To determine the lunar parallax he resorted to direct observations of the zenith distance of the Moon on the meridian, comparing the result of his observations with the position obtained from the lunar theory. He determined the parallax when the Moon was nearest the zenith, and also when it crossed his meridian at its farthest distance from the zenith. From his observations he obtained results varying from less than 50 per cent of the true parallax (57′.0) to more than 150 per cent of that value. According to Houzeau the definitive result of Ptolemy's work is 58′.7.

It is thus seen that the astronomers of two thousand years ago had a fairly accurate knowledge of the distance of the Moon from the Earth, but an entirely erroneous one of the distance of the Sun, the true distance being something like twenty times that assumed by them. This value of the distance of the Sun from the Earth was accepted for nineteen centuries from Aristarchus to Kepler, having been deduced anew by such men as Copernicus and Tycho Brahe.

With the announcement by Kepler, early in the seventeenth century, of his laws of planetary motion, it became possible to deduce from the periodic times of revolution of the planets around the Sun their relative distances from that body, and thus to determine the distance of the Sun from the Earth, by determining the distance or parallax of one of the planets.

From observations of Mars, Kepler obtained the distance of the Sun from the Earth as about three times that accepted up to his time. His value, however, was but one-seventh of the true distance. About fifty years later Flamsteed and Cassini working independently and using the same method as that employed by Kepler obtained for the first time approximately the correct value of the distance of the Sun from the Earth. In a letter, dated November 16, 1672, to the Publisher of the *Philosophical Transactions*, Flamsteed says:

> September last I went to Townley. The first week that I intended to have observed ♂ there with Mr. Townley, I twice observ'd him, but could not make two Observations, as I intended, in one night. The first night after my return, I had the good hap to measure his distances from two Stars the same night; whereby I find, that the Parallax was very small: certainly not 30 seconds: So that I believe the Sun's Parallax is not more than 10 seconds. Of this Observation I intend to write a small Tract, when I shall gain leisure; in which I shall demonstrate both the Diameter and Distances of all the Planets by Observations; for which I am now pretty well fitted.

During the two and a half centuries since Flamsteed's determination there have been more than a hundred determinations of the solar parallax by various methods. In the method used by Flamsteed, the rotation of the Earth is depended upon to change the relative position of the observer, the center of the Earth, and Mars. (Diagram shown) Another method is to establish two stations widely separated in latitude, and in approximately the same longitude. At one station, the zenith distance of Mars will be determined as it crosses the meridian north of the zenith; at the other station, the zenith distance will be determined as it crosses the meridian south of the zenith. The sum of the two zenith distances minus the difference in latitude between the two stations will give the displacement of Mars due to parallax. These two methods have been successfully applied to several of the asteroids whose distances from the Sun are very nearly that of Mars.

The nearest approach of Venus to the Earth is during her transit across the face of the Sun, and these occasions, four during the last two centuries, have been utilized to determine the solar parallax. Here as in the case of Mars two different methods may be used, either by combining observations at two stations widely separated in latitude, or at two stations widely separated in longitude. (Diagrams shown.)

The methods just described for obtaining the solar parallax, the geometrical methods, were made available, as has been said, by the discovery of Kepler's laws of planetary motion. Newton's discovery of the law of gravitation gave rise to another group of methods, designated as gravitational methods. The best of these is probably that in which the distance of the Sun from the Earth is determined from the mass of the Earth, which, in turn, is determined from the perturbative effect of the Earth upon Venus and Mars. This method is long and laborious but its importance lies in the fact that the accuracy of the result increases with the time. Professor C. A. Young says:

> this is the "method of the future," and two or three hundred years hence will have superseded all the others,—unless indeed it should appear that bodies at present unknown are interfering with the movements of our neighboring planets, or unless it should turn out that the law of gravitation is not quite so simple as it is now supposed to be.

A third group of methods of determining the distance of the Sun from the Earth, called the physical methods, depends upon the determination of the velocity of light in conjunction either with the time it takes light to travel from the Sun to the Earth obtained from observations of the eclipses of Jupiter's satellites, or with the constant of aberration derived from observations of the stars.

In August, 1898, Dr. Witt of Berlin discovered an asteroid, since named Eros, which was soon seen to offer exceptional opportunity for the determination of the solar parallax, as at the very next opposition, in November, 1900, it would approach to within 30,000,000 miles of the Earth. At the meeting of the Astrographic Chart Congress in Paris in July, 1900, it was resolved to seize this opportunity and organize an international parallax campaign. Fifty-eight observatories took part in the various observations called for by the general plan. The meridian instruments determined the absolute position of Eros from night to night as it crossed the meridians of the various observatories; the large visual refractors measured the distance of Eros, from the faint stars near it, at times continuing the measures throughout the entire night; and the photographic equatorial:, obtained permanent records of the position of Eros among the surrounding stars. In addition long series of observations had to be made to determine the positions of the stars to which Eros was referred.

When several years had elapsed after the completion of the observations, and no discussion of all the material had been provided for, Prof. Arthur R. Hinks of Cambridge, England volunteered for the work. The undertaking was truly monumental. He first formed a catalogue of the 671 stars which had been selected by the Paris Congress for observation as marking out the path of Eros from a discussion of the results obtained by the meridian instruments and from the photographic plates. This done, with these results as a basis, a larger catalogue of about 6000 stars had to be formed from measures on the photographic plates. He was then ready to commence the discussion of the observations of Eros itself. From 1901 to 1910 there appeared in the *Monthly Notices of the Royal Astronomical Society* eight articles covering 135 pages giving the results of his labors.

From a discussion of all the photographic observations he obtained a solar parallax of

$$8''.807 \pm 0''.0027$$

a probable error equivalent to an uncertainty of about 30,000 miles in the distance to the Sun.

From a discussion of all the micrometric observations he obtained

$$8''.806 \pm 0''.004$$

The observations with the meridian instruments gave

$$8''.837 \pm 0''.0185$$

a determination relatively much weaker than either of the others.

A parallax of 8″.80, the value adopted for all the national almanacs twenty years ago, corresponds to a distance of 92,900,000 miles. At present it seems improbable that another parallax campaign will be undertaken before 1931, when Eros approaches still nearer to the Earth, its least distance at that time being about 15,000,000 miles.

Table I
Approximate Distance from Earth to Sun as Accepted at Various Times

DATE	DISTANCE
	miles
275 B.C. to 1620 A.D.	4,500,000
1620 Kepler	13,500,000
1672 Flamsteed	81,500,000
1916	92,900,000

When Copernicus proposed that the Sun is the center of the Solar System and that all the planets including the Earth revolve around the Sun, it was at once seen that such a motion of the Earth must produce an annual parallax of the stars. Tycho Brahe rejected the Copernican System because he could not find from his observations any such parallax. However, the system was generally accepted as the true one and the determination of stellar parallax or the distance of the stars became a live subject. Picard in the latter half of the seventeenth century, using a telescope and a micrometer in connection with his divided circle, showed an annual variation in the declination of the pole star amounting to 40″. In 1674 Hooke announced a parallax of 15″ for γ Draconis. About this same time Flamsteed announced a parallax of 20″ for α Ursae Minoris, but J. Cassini showed that the variations in the declination did not follow the law of the parallax.

The period which we have now reached is so admirably treated by Sir Frank W. Dyson, Astronomer Royal, in his Halley Lecture delivered at Oxford on May 20, 1915, that I ask your indulgence while I quote rather freely from that source.

Thus in Halley's time, it was fairly well established that the stars were at least 20,000 or 30,000 times as distant as the sun. Halley did not succeed in finding their range, but he made an important discovery which showed that three of the stars were at sensible distances. In 1718 he contributed to the Royal Society a paper entitled Considerations of the Change of the Latitude of Some of the Principle Bright Stars. While pursuing researches on another subject, he found that the three bright stars—Aldebaran, Sirius, and Arcturus—occupied positions among the other stars differing considerably from those assigned to them in the Almagest of Ptolemy. He showed that the possibility of an error in the transcription of the manuscript could be safely excluded, and that the southward movement of these stars to the extent of 37′, 42′, and 33′,—*i.e.* angles larger than the apparent diameter of the sun in the sky—were established. . . .

This is the first good evidence, *i.e.* evidence which we now know to be true, that the so called fixed stars are not fixed relatively to one another. It is the first positive proof that the distances of the stars are sensibly less than infinite.

At the time of the appearance of Halley's paper there was coming into notice a young astronomer, James Bradley, then 26 years old. He was admitted to membership in the Royal Society the same year that Halley's paper was presented. He was exceedingly eager to attack the problem of the distances of the stars. At length the opportunity presented itself. To quote again from Sir Frank Dyson:

Bradley designed an instrument for measuring the angular distance from the zenith, at which a certain star, γ Draconis, crossed the meridian. This instrument is called a zenith sector. The direction of the vertical is given by a plumb-line, and he measured from day to day the angular distance of the star from the direction of the vertical. From December, 1725, to March, 1726, the star gradually moved further south; then it remained stationary for a little time; then moved northwards until, by the middle of June, it was in the same position as in December. It continued to move northwards until the beginning of September, then turned again and reached its old position in December. The movement was very regular and evidently not due to any errors in Bradley's observations. But it was most unexpected. The effect of parallax—which Bradley was looking for—would have brought the star farthest south in December, not in March. The times were all three months wrong. Bradley examined other stars thinking first that this might be due to a movement of the Earth's pole. But this would not explain the phenomena. The true explanation, it is said, although I do not know how truly, occurred to Bradley when he was sailing on the Thames, and noticed that the direction of the wind, as indicated by a vane on the mast-head, varied slightly with the course on which the boat was sailing. An account of the observations in the form of a letter from Bradley to Halley is published in the *Philosophical Transactions* for December, 1728:

> When the year was completed, I began to examine and compare my observations, and having pretty well satisfied myself as to the general laws of the phenomena, I then endeavored to find out the cause of them. I was already convinced that the apparent motion of the stars was not owing to the nutation of the earth's axis. The next thing that offered itself was an alteration in the direction of the plumb-line with which the instrument was constantly rectified; but this upon trial proved insufficient. Then I considered what refraction might do, but there also nothing satisfactory occurred. At length I conjectured that all the phenomena hitherto mentioned, proceeded from the progressive motion of light and the earth's annual motion in its orbit. For I perceived that, if light was propagated in time, the apparent place of a fixed object would not be the same when the eye is at rest, as when it is moving in any other direction than that of the line passing through the eye and the object; and that, when the eye is moving in different directions, the apparent place of the object would be different.

When Bradley's observations of γ Draconis were corrected for aberration, they showed, according to himself, that the parallax of that star could not be as much as $1''.0$, or that the star was more than 200,000 times as distant from the Earth as the Sun.

On December 6, 1781 there was read before the Royal Society a paper by Mr. Herschel, afterwards Sir William, on the Parallax of the Fixed Stars. We read:

The method pointed out by Galileo, and first attempted by Hook, Flamstead, Molineaux, and Bradley, of taking distances of stars from the zenith that pass very near it, though it failed with regard to parallax, has been productive of the most noble discoveries of another nature. At the same time it has given us a much juster [sic] idea of the immense distance of the stars, and furnished us with an approximation to the knowledge of their parallax that is much nearer the truth than we ever had before. . . .

In general, the method of zenith distances labours under the following considerable difficulties. In the first place, all these distances, though they should not exceed a few degrees, are liable to refractions; and I hope to be pardoned when I say that the real quantities of these refractions, and their differences, are very far from being perfectly known. Secondly, the change of position of the earth's axis arising from nutation, precession of the equinoxes, and other causes, is so far from being completely settled, that it would not be very easy to say what it exactly is at any given time. In the third place, the aberration of light, though best known of all, may also be liable to some small errors, since the observations from which it was deduced laboured under all the foregoing difficulties. I do not mean to say, that our theories of all these causes of error are defective; on the contrary, I grant that we are for most astronomical purposes sufficiently furnished with excellent tables to correct our observations from the above mentioned errors. But when we are upon so delicate a point as the parallax of the stars; when we are investigating angles that may, perhaps, not amount to a single second, we must endeavour to keep clear of every possibility of being involved in uncertainties; even the hundredth part of a second becomes a quantity to be taken into consideration.

Herschel then proceeds to advocate selecting pairs of stars of very unequal magnitude and whose distance apart is less than five seconds, and making very accurate micrometric measures of this distance from time to time. The first condition, should give, in general, stars very unequally distant from the Earth, so that the changing perspective as the Earth revolves in her orbit would give a variation of the apparent distance between the stars, while the small distance, less than five seconds, would eliminate from consideration entirely any effect upon this distance of the uncertainties in refraction, precession, nutation, aberration, *etc*. Herschel had already commenced the cataloguing of such double stars and in January, 1782, submitted to the Royal Society a catalogue of 269. This work did not enable Herschel to determine the distances of the stars but did enable him to demonstrate that there exist pairs of stars in which the two components revolve the one around the other. In twenty years he had found fifty such pairs.

Coming forward another generation, that is, to a time a little less than a hundred years ago, we find Pond, then Astronomer Royal, writing:

The history of annual parallax appears to me to be this: in proportion as instruments have been imperfect in their construction, they have misled observers into the belief of the existence of sensible parallax. This has happened in Italy to astronomers of the very first reputation. The Dublin instrument is superior to any of a similar construction on the continent; and accordingly it shows a much less parallax than the Italian astronomers imagined they had detected. Conceiving that I have established, beyond a doubt, that the Greenwich instrument approaches still nearer to perfection, I can come to no other conclusion than that this is the reason why it discovers no parallax at all.

Within fifteen years after this statement by Pond, observations had been obtained which showed a measurable parallax of three different stars. The announcements of these results, each by a different astronomer, were practically simultaneous.

W. Struve, using a filar micrometer, determined the distance of α Lyrae from a small star about 40″ distant on 60 different days over a period of nearly three years. He obtained a parallax of 0″.262 ± 0″.025. Bessel, using his heliometer, determined the distances of 61 Cygni from two small stars distant about 500″ and 700″ respectively. He obtained for this star a parallax of 0″.314 ± 0″.020. Henderson, using determinations of the position of α Centauri by meridian instruments, deduced a parallax of 1″.16 ± 0″.11. All three of these results were announced in the winter of 1838-39, and indicate that the three stars are distant from the Earth about 750,000, 650,000, and 200,000 times the distance of the Sun from the Earth.

Table II exhibits the observed displacement of 61 Cygni by monthly means as given by Main from Bessel's observations. The last column gives the computed displacement on the assumption of a parallax of 0″.314. The reality of the parallax is seen at a glance.

In 1888, fifty years after the first determination of what we now know to be a true stellar parallax, Young in his *General Astronomy* gives, in a list of known stellar parallaxes, 28 stars and 55 separate determinations. Within the next ten years the number of stars whose parallaxes had been determined about doubled, due principally to the work of Gill and Elkin.

TABLE II
Parallax of 61 Cygni

	Observed Displacement	Computed from 0″.314
1837 August 23	+0″.20	+0.18
September 14	+0.10	+0.08
October 12	+0.04	−0.05
November 22	−0.21	−0.22
December 21	−0.32	−0.27
1838 January 14	−0.38	−0.27
February 5	−0.22	−0.23
May 14	+0.24	+0.20
June 19	+0.36	+0.28
July 13	+0.22	+0.28
August 19	+0.15	+0.19
September 19	+0.04	+0.06

Probably the most extensive piece of stellar parallax work in existence is that with the Yale heliometer. The results to date were published in 1912 and contained the parallaxes of 245 stars, the observations extending over a quarter of a century, the entire work having been done by three men, Elkin, Chase, and Smith. In selecting a list of stars for parallax work an effort is made to obtain stars which give promise of being nearer than the mass of stars. At first the brighter stars were selected, and then those with large proper motions. The Yale list of 245 stars contains all stars in the northern heavens whose annual proper motion is known to be as much as 0″.5. Of these 245 stars, 54 are given a negative parallax. A negative parallax does not mean, as someone has expressed it, that the star is "somewhere on the other side of nowhere," but such a result may be attributed to the errors of observation or to the fact that the comparison stars are nearer than the one under investigation. It is safe to say, however, that somewhat more than half of the 245 stars have a measurable parallax.

Another series of stellar parallax observations, comparable in extent with the one just mentioned, is that of Flint at the Washburn Observatory. This series includes 203 stars and extended from 1893 to 1905. These observations were made with a meridian circle, but not after the method of a century ago. The observations were strictly differential, the general plan being to select two faint comparison stars, one immediately preceding and the other immediately following the parallax star, and to determine the difference in right ascension, the observation of the three stars occupying about five minutes. Here as in the case of the Yale heliometer work a large proportion of the resulting parallaxes are negative; somewhat more than half, however, were found to have a measurable parallax. The average probable error of a parallax was the same in each of these two pieces of work, about 0″.03. The progress of the work during the last two or three generations is given in Table III which contains also a brief statement of the discoveries made during the preceding century due chiefly to efforts to measure stellar parallaxes.

TABLE III
Approximate Number of Known Stellar Parallaxes

DATE.	ASTRONOMER	NUMBER OF STARS WITH KNOWN PARALLAXES	DISCOVERIES
1718	Halley	No parallax.	Proper motion.
1728	Bradley	No parallax.	Aberration.
1750	Bradley	No parallax.	Nutation.
1790	Herschel	No parallax.	True binary systems.
1838		3.	
1888		28.	
1898		50 to 60.	
1916		200 to 300.	

A generation ago photography entered the field of stellar parallax work, and has outdistanced all the previously employed methods for efficiency. In 1911, two publications appeared giving the results of photographic stellar parallax work, one by Russell, giving the parallaxes of forty stars from photographs taken by Hinks and himself at Cambridge, England, the other by Schlesinger, giving the parallaxes of twenty-five stars from photographs taken mostly by himself at the Yerkes Observatory, Williams Bay, Wisconsin. In speaking of these two series of observations, Sir David Gill said:

On the whole, the Cambridge results, when a sufficient number of plates have been taken, and when the comparison stars are symmetrically arranged, give results of an accuracy which, but for the wonderful precision of the Yerkes observations, would have been regarded as of the highest class.

Schlesinger has shown that with a telescope of the size and character of the Yerkes instrument "the number of stellar parallaxes that can be determined per annum, with an average probable error of 0″.013, will in the long run be about equal to the number of clear nights available for the work."

In other words, the Yerkes 40-inch equatorial used photographically determines stellar parallaxes with one-tenth the labor required with an heliometer and with twice the accuracy.

In July, 1913, stellar parallax work was undertaken with the 60-inch reflector of the Mount Wilson Solar Observatory, and at the meeting of the American Astronomical Society at San Francisco in August, 1915 a report on that work was made. The parallaxes of thirteen stars had been determined, with a maximum probable error of 0″.010 and an average probable error of less than 0″.006, giving twice the accuracy of the Schlesinger results with the Yerkes 40-inch and from three to five times that obtained fifteen years ago. What may we not expect when the 100-inch reflector gets to work on Mt. Wilson.

At the meeting of the American Astronomical Society to which reference has just been made, two other observatories reported upon their stellar parallax work. Lee and Joy of the Yerkes Observatory reported the parallaxes of nine stars with a maximum probable error of 0″.014 and an average probable error of 0″.010, and Mitchell of Leander McCormick Observatory reported the parallaxes of eleven stars with a maximum probable error of 0″.012 and an average probable error of 0″.009.

The progress made in the accuracy of parallax results is shown at a glance in Table IV.

From these results it appears that any star whose parallax is as much as 0″.02, *i.e.*, whose distance from the Earth is less than ten million times that from the Earth to the Sun, should give a positive result when subjected to the treatment now employed in parallax investigations, and as eight or ten observatories are devoting their energies to stellar parallax work at present, the combined programs containing over 1000 different stars, we ought soon to have lists of at least a few thousand stars whose parallaxes are known where our present lists contain but a few hundred.

TABLE IV
The Accuracy of Stellar Parallax Determinations

Date		Instrument	Probable error	Observers
1838	Micrometric	Dorpat Refractor	$0''.025$	Struve.
1838		Konigsberg heliometer	0.02	Bessel.
1880-1898		Cape heliometer	0.017	Gill and Assistants.
1888-1912		Yale heliometer	0.03	Elkin, Chase, and Smith.
1893-1905		Washburn meridian circle	0.03	Flint.
1910	Photograph	Yerkes Refractor	0.013	Schlesinger.
1915		Yerkes Refractor	0.010	Lee and Joy.
1915		Leander McCormick Refractor	0.009	Mitchell.
1915		Mt. Wilson 60 inch Reflector	0.006	Van Maanan.

31

Introduction to

Modern Theories of the Spiral Nebulae

Heber D. Curtis (1872 – 1942) was an astronomer who worked at Lick Observatory in Mt. Hamilton, California and at Allegheny Observatory in Pittsburgh. He is, perhaps, most well-known for the 1920 Shapley-Curtis debate on the distances to galaxies (spiral nebulae). Although Shapley won the debate, it was Curtis who correctly placed "spiral nebulae" outside the Milky Way. The following article appeared in 1918 in the Washington Academy of Sciences Journal Volume 9.

Today we know that spiral nebulae are distant galaxies thanks to the work of astronomers like Edwin Hubble. At the time of this article, however, there were few indications of their great distances. Here Curtis pulls together several compelling pieces of evidence that spiral nebulae were indeed 'island universes' and outside the expanse of the Milky Way thus setting the stage for later confirming work by Edwin Hubble and others.

Modern Theories of the Spiral Nebulae[1]

HEBER D. CURTIS

Lick Observatory
(Communicated by W. J. Humphreys)

In one sense, that theory of the spiral nebulae to which many lines of recently obtained evidence are pointing, cannot be said to be a modern theory. There are few modern concepts which have not been explicitly or implicitly put forward as hypotheses or suggestions long before they were actually substantiated by evidence.

The history of scientific discovery affords many instances where men with some strange gift of intuition have looked ahead from meager data, and have glimpsed or guessed truths which have been fully verified only after the lapse of decades or centuries. Herschel was such a fortunate genius. From the proper motions of a very few stars he determined the direction of the sun's movement nearly as accurately, due to a very happy selection of stars for the purpose, as far more elaborate modern investigations. He noticed that the star clusters which appeared nebulous in texture in smaller telescopes and with lower powers, were resolved into stars with larger instruments and higher powers. From this he argued that all the nebulae could be resolved into stars by the application of sufficient magnifying power, and that the nebulae were, in effect, separate universes, a theory which had been earlier suggested on purely hypothetical or philosophical grounds, by Wright, Lambert, and Kant. From their appearance in the telescope he, again with almost uncanny prescience, excepted a few as definitely gaseous and irresolvable.

This view held sway for many years; then came the results of spectroscopic analysis showing that many nebulae (those which we now classify as diffuse or planetary) are of gaseous constitution and cannot be resolved into stars. The spiral nebulae, although showing a different type of spectrum, were in most theories tacitly included with the known gaseous nebulae.

---

[1] Abstract of a lecture given on March 15, 1918, at a joint meeting of the Washington Academy of Sciences and the Philosophical Society of Washington. The lecture was illustrated with numerous lantern slides.

We have now, as far as the spiral nebulae are concerned, come back to the standpoint of Herschel's fortunate, though not fully warranted deduction, and the theory to which much recent evidence is pointing, is that these beautiful objects are separate galaxies, or "island universes," to employ the expressive and appropriate phrase coined by Humboldt.

By means of direct observations on the nearer and brighter stars, and by the application of statistical methods to large groups of the fainter or more remote stars, the galaxy of stars which forms our own stellar universe is believed to comprise perhaps a billion suns. Our sun, a relatively inconspicuous unit, is situated near the center of figure of this galaxy. This galaxy is not even approximately spherical in contour, but shaped like a lens or thin watch; the actual dimensions are highly uncertain; Newcomb's estimate that this galactic disk is about 3,000 light-years in thickness, and 30,000 light-years in diameter, is perhaps as reliable as any other.

Of the three classes of nebulae observed, two, the diffuse nebulosities and the planetary nebulae, are typically a galactic phenomenon as regards their apparent distribution in space, and are rarely found at any distance from the plane of our Milky Way. With the exception of certain diffuse nebulosities whose light is apparently a reflection phenomenon from bright stars involved within the nebulae, both these types are of gaseous constitution, showing a characteristic bright-line spectrum.

Differing radically from the galactic gaseous nebulae in form and distribution, we find a very large number of nebulae predominantly spiral in structure. The following salient points must be taken into account in any adequate theory of the spiral nebulae.

1. In apparent size the spirals range from minute flecks, just distinguishable on the photographic plate, to enormous spirals like *Messier* 33 and the Great Nebula in Andromeda, the latter of which covers an area four times greater than that subtended by the full moon.
2. Prior to the application of photographic methods, fewer than ten thousand nebulae of all classes had been observed visually. One of the first results deduced by Director Keeler from the program of nebular photography which he inaugurated with the Crossley Reflector at Lick Observatory, was the fact that great numbers of small spirals are within reach of modern powerful reflecting telescopes. He estimated their total number as 120,000 early in the course of this program, and before plates of many regions were available. I have recently made a

count of the small nebulae on all available regions taken at the Lick Observatory during the past twenty years[2] and from these counts estimate that there are at least 700,000 spiral nebulae accessible with large reflectors.

3. The most anomalous and inexplicable feature of the spiral-nebulae is found in their peculiar distribution. They show an apparent abhorrence for our galaxy of stars, being found in greatest numbers around the poles of our galaxy. In my counts I found an approximate density of distribution as follows:

Galactic Latitude +45° to + 90°   34 per square degree.
Galactic Latitude −45° to −90°   28 per square degree.
Galactic Latitude ±30° to ±45°   24 per square degree.
Galactic Latitude −30° to +30°   7 per square degree.

4. No spiral has as yet been found actually within the structure of the Milky Way. We have doubled and trebled our exposures in regions near the galactic plane in the hope of finding fainter spirals in such areas, but thus far without results. The outstanding feature of the space distribution of the spirals is, then, that they are found in greatest profusion where the stars are fewest, and do not occur where the stars are most numerous. This distribution may be illustrated graphically as follows:[3]

THE FACTORS OF SPACE DISTRIBUTION

400,000 Spiral Nebulae

..
..
..
..

Our own stellar universe is shaped like a thin lens,
and is perhaps 3,000 by 30,000 light-years in extent. In this
space occur nearly all the stars, nearly all the new stars, nearly all
the variable stars and most of the diffuse and planetary
nebulae, etc., but *no spiral nebulae.*

..
..
..
..

300,000 Spiral Nebulae

---

[2] CURTIS, H. D. *On the number of spiral nebulae,* Proc. Amer. Phil. Soc. 57: 513. 1918
[3] [Ed.] Using words the author lays out the distribution of celestial objects as seen from the side. The Milky Way appears in the center as an oblate disk.

The spectrum of the spirals is practically the same as that given by a star cluster, showing a continuous spectrum broken by absorption lines. A few spirals show bright-line spectra in addition.

5. The space-velocities of the various classes of celestial objects are summarized in the following short table:

TABLE
THE FACTORS OF SPACE-VELOCITY

1. *The Diffuse Nebulae*
   Velocities low.
2. *The Stars*
   Velocities vary with spectral type.
   Class B Stars: average speeds 8 miles per second.
   Class A Stars: average speeds 14 miles per second.
   Class F Stars: average speeds 18 miles per second.
   Class G Stars: average speeds 19 miles per second.
   Class K Stars: average speeds 21 miles per second.
   Class M Stars: average speeds 21 miles per second.
3. *The Star Clusters*
   Velocities unknown.
4. *The Planetary Nebulae*
   Average speeds 48 miles per second.
5. *The Spiral Nebulae*
   Average speeds 480 miles per second.

The peculiar variation of the space-velocity of the stars with spectral type may ultimately prove to be a function of relative mass. The radial velocities of but few spirals have been determined to date; future work may change the value given, but it seems certain that it will remain very high.

It will be seen at once that, with regard to this important criterion of space-velocity, the spiral nebulae are very distinctly in a class apart. It seems impossible to place them at any point in a coherent scheme of stellar evolution. We cannot bridge the gap involved in postulating bodies of such enormous space velocities either as a point of stellar origin, or as a final evolution product.

On the older theory that the spirals are a part of our own galaxy, it is impossible to harmonize certain features of the data thus far presented.

If this theory is true, their grouping near the galactic poles, inasmuch as all evidence points to a flattened or disk form for our galaxy, would indicate that they are relatively close to us. In that event, we should inevitably have detected in this class of objects proper motions of the same order of magnitude as those found for the stars at corresponding distances. Such proper motions are the more to be expected in view of the fact that the average space velocity of the spirals is about thirty times that of the stars. I have repeated all the earlier plates of the Keeler nebular program, and was able to find no certain evidence of either translation or rotation in these objects in an average time interval of thirteen years.[4] Their form, and the evidence of the spectroscope, indicate, however, that they are in rotation. Knowing that their space-velocities are high, the failure to detect any certain evidence of cross motion is an indication that these objects must be very remote.

Even if the spiral is not a stage in stellar evolution, but a class apart, is it still possible to assume that they are, notwithstanding, an integral part of our own stellar universe, sporadic manifestations of an unknown line of evolutionary development, driven off in some mysterious manner from the regions of greatest star density?

A relationship between two classes of objects may be one of avoidance just as logically as one of contiguity. It has been argued that the absolute avoidance which the spirals manifest for the galaxy of the stars shows incontrovertibly that they must, by reason of this very relationship of avoidance, be an integral feature of our galaxy. This argument has proved irresistible to many, among others to so keen a thinker as Herbert Spencer, who wrote:

> In that zone of celestial space where stars are excessively abundant nebulae are rare; while in the two opposite celestial spaces that are furthest removed from this zone nebulae are abundant . . . . . . Can this be mere coincidence? When to the fact that the general mass of the nebulae are antithetical in position to the general mass of the stars, we add the fact that local regions of nebulae are regions where stars are scarce . . . . does not the proof of a physical connection become overwhelming?

---

[4] CURTIS, H. D. *The proper motion of the nebulae.* Publ. Astron. Soc. Pacific 27: 214. 1915.

It must be admitted that a distribution, which has placed three-quarters of a million objects around the poles of our galaxy, would be against all probability for a class of objects which would be expected to be arranged at random, unless it can be shown that this peculiar grouping is only apparent, and due to some phenomenon in our own galaxy. This point will be reverted to later.

It has been shown that the factors of space-velocity and space-distribution separate the spirals very clearly from the stars of our galaxy; from these facts alone, and from the evidence of the spectroscope, the island universe theory is given a certain measure of credibility.

Another line of evidence has been developed within the past two years, which adds further support to the island-universe theory of the spiral nebulae.

NEW STARS

Within historical times some twenty-seven new stars have suddenly flashed out in the heavens. Some have been of interest only to the astronomer; others, like that of last June, have rivaled *Sirius* in brilliancy. All have shown the same general history, suddenly increasing in light ten thousand-fold or more, and then gradually, but still relatively rapidly, sinking into obscurity again. They are a very interesting class, nor has astronomy as yet been able to give any universally accepted explanation of these anomalous objects. Two of these novae had appeared in spiral nebulae, but this fact had not been weighed at its true value. Within the past two years over a dozen novae have been found in spiral nebulae, all of them very faint, ranging from about the fourteenth to the nineteenth magnitudes at maximum. Their life history, so far as we can tell from such faint objects, appears to be identical with that of the brighter novae. Now the brighter novae of the past, that is, those which have not appeared in spirals, have almost invariably been a galactic phenomenon, located in or close to our Milky Way, and they have very evidently been a part of our own stellar system. The cogency of the argument will, I think, be apparent to all, although the strong analogy is by no means a rigid proof. If twenty-seven novae have appeared in our own galaxy within the past three hundred years, and if about half that number are found within a few years in spiral nebulae far removed from the galactic plane, the presumption that these spirals are themselves galaxies composed of hundreds of millions of stars is a very probable one.

If, moreover, we make the reasonable assumption that the new stars in the spirals and the new stars in our own galaxy average about the same in size, mass, and absolute brightness, we can form a very good estimate of the probable distance of the spiral nebulae, regarded as island universes. Our galactic novae have averaged about the fifth magnitude. The new stars which have appeared in the spiral nebulae have averaged about the fifteenth magnitude, but it would appear probable that we must inevitably miss the fainter novae in such distant galaxies, and it is perhaps reasonable to assume that the average magnitude of the novae in spirals may be about the eighteenth, or thirteen magnitudes fainter than those in our own galaxy. They would thus be about 160,000 times fainter than our galactic novae, and on the assumption that both types of novae average the same in mass, absolute luminosity, *etc.*, the novae in spirals should be four hundred times further away. We do not know the average distance of the new stars which have appeared in our own galaxy, but 100,000 light-years is perhaps a reasonable estimate. This would indicate a distance of the order of 4,000,000 light-years for the spiral nebulae. This is an enormous distance, but, if these objects are galaxies like our own stellar system, such a distance accords well with their apparent dimensions. Our own galaxy, at a distance of 10,000,000 light-years, would be about 10 minutes of arc in diameter, or the size of the larger spiral nebulae.

On such a theory, a spiral structure for our own galaxy would be probable. Its proportions accord well with the degree of flattening observed in the majority of the spirals. We have very little actual evidence as to a spiral structure for our galaxy; the position of our sun relatively close to the center of figure of the galaxy, and our ignorance of the distances of the remoter stars, renders such evidence very difficult to obtain. A careful study of the configurations and star densities in the Milky Way has led Professor Easton, of Amsterdam, to postulate a spiral structure for our galaxy.

DISTRIBUTION OF SPIRALS

There is still left one outstanding and unexplained problem in the island universe theory or any other theory of the spiral nebulae. Neither theory, as outlined, offers any satisfactory explanation of the remarkable distribution of the spirals. On the older theory, if a feature of our galaxy, what has driven them out to the points most remote from the regions of greatest star density? If, on the other hand, the spirals are island universes, it is against all probability that our own universe should have chanced to be situated about half way between two great groups of island universes, and that not a single object of the class happens to be located in the plane of our Milky Way.

There is one very common characteristic of the spirals which may be tentatively advanced as an explanation of the peculiar grouping of the spirals.

A very considerable proportion of the spirals show indubitable evidence of occulating matter, lying in the plane of the greatest extension of the spiral, generally outside the whorls, but occasionally between the whorls as well. This outer ring of occulting matter is most easily seen when the spiral is so oriented in space as to turn its edge toward us. But the phenomenon is also seen in spirals whose planes make a small, but appreciable angle with our line of sight, manifesting itself in such appearances as "lanes" more prominent on one side of the major axis of the elongated elliptical projection, in a greater brightness of the nebular matter on one side of this major axis, in a fan-shaped nuclear portion, or in various combinations of these effects. The phenomenon is a very common one. Illustrations of seventy-eight spirals showing evidences of occulting matter in their peripheral equatorial regions, with a more detailed discussion of the forms observed, are now being published,[5] and additional examples of the phenomenon are constantly being found.

While we have as yet no definite proof of the existence of such a ring of occulting matter lying in our galactic plane and outside of the great mass of the stars of our galaxy, there is a great deal of evidence for such occulting matter in smaller areas in our galaxy. Many such dark areas are observed around certain of the diffuse nebulosities, or seen in projection on the background furnished by such nebulosities or the denser portions of the Milky Way; these appearances seem to be actual "dark nebulae."[6] The curious "rifts" in the Milky Way may well be ascribed, at least in part, to such occulting matter.

Though we thereby run the risk of arguing in a circle, the fact that no spirals can be detected in our galactic plane, a natural result of such a ring of occulting matter, would in itself appear to lend some probability to the hypothesis. The peculiar distribution of the spiral nebulae would then

---

[5] CURTIS, H. D. *Occulting effects of spiral nebulae.* Univ. Calif. Semi-Cent. Publ. (in press).

[6] BARNARD, E. E. On *the dark markings of the sky,* with a catalogue of 182 such objects. Astrophys. Journ. **49: 1.** 1919; CURTIS, H. D. *Dark nebulae.* Publ. Astron. Soc. Pacific 30: 65. 1918.

be explained as due, not to an actual asymmetrical and improbable distribution in space, but to a cause within our own galaxy, assumed to be a spiral with a peripheral ring of occulting matter similar to that observed in a large proportion of the spirals. The argument that the spirals must be an integral feature of our own galaxy, based on a relationship of avoidance, would then lose its force. The explanation appears to be a possibility, even a strong probability, on the island universe theory, and I know of no other explanation, on any theory, for the observed phenomenon of nebular distribution about our galactic poles.

SUMMARY

The Spiral Nebulae as Island Universes

1. On this theory, it is unnecessary to attempt to coordinate the tremendous space-velocities of the spirals with the thirty-fold smaller values found for the stars. Very high velocities have been found for the Magellanic Clouds, which may possibly be very irregular spirals, relatively close to our galaxy.
2. There is some evidence for a spiral structure in our own galaxy.
3. The spectrum of the majority of the spirals is practically identical with that given by a star cluster; a spectrum of this general type is such as would be expected from a vast congeries of stars.
4. If the spirals are separate universes, similar to our galaxy in extent and in number of component stars, we should observe many new stars in the spirals, closely resembling in their life history the twenty-seven novae which have appeared in our own galaxy. Over a dozen such novae in spirals have been found, and it is probable that a systematic program of repetition of nebular photographs will add greatly to this number. A comparison of the average magnitudes of the novae in spirals with those of our own galaxy indicates a distance of the order of 10,000,000 light-years for the spirals. Our own galaxy at this distance would appear 10 minutes in diameter, the size of the larger spirals.
5. A considerable proportion of the spirals show a peripheral equatorial ring of occulting matter. So many instances of this have been found that it appears to be a general though not universal characteristic of the spirals; the existence of such an outer ring of occulting matter in our own galaxy, regarded as a spiral, would furnish an adequate explanation of the peculiar distribution of the spirals. There is considerable evidence of such occulting matter in our galaxy.

An English physicist has cleverly said that any really good theory brings with it more problems than it removes. It is thus with the island-universe theory. It is impossible to do more than to mention a few of these problems, with no attempt to divine those which may ultimately be presented to us.

While the data are too meager as yet, several attempts have been made to deduce the velocity of our own galaxy within the super-galaxy. It would not be surprising if the space-velocity of our galaxy, like those of the spirals and the Magellanic Clouds, should prove to be very great, hundreds of miles per second.

Further, what are the laws which govern the forms assumed, and under which these spiral whorls are shaped? Are they stable structures; are the component stars moving inward or outward? A beginning has been made by Jeans and other mathematicians on the dynamical problems involved in the structure of the spirals. The field for research is, like our subject matter, practically infinite.

Introduction to

Exact Time in Astronomy

Jean Boccardi was an active researcher in the rotation of the planets, the figure of the Earth, variations of latitude, exact positions of stars, and determination of time. The following article appeared in 1928 in the Washington Academy of Sciences Journal Volume 18.

In this article Boccardi discusses the errors inherent in measuring time with a mechanical clock (a Riefler clock) and a "broken telescope" to determine the exact moment of stellar transit across the local meridian. The stellar transit method was commonly used to set the length of the sidereal day. The astronomer would carefully measure the time at which a set of standard stars would cross the local meridian (as determined by a thin cross wire in the telescope eyepiece). Today this work is done using very long baseline interferometry. "Broken" telescopes are easily reversed – the telescope can be quickly rotated around its horizontal axis. They are compact, and the viewer eyepiece remains at one level throughout the observation.

Exact Time in Astronomy

Jean Boccardi,

Varazze, Liguria, Italy

I

For about a century, as a result of the establishment of the principles and rules of the theory of errors, it has been the custom in sciences of observation and measurement to give, along with the numerical values obtained, the probable error, or the mean square error, which this theory enables us to assign. Experience has shown that the more closely the conditions of observation approximate to the theoretical conditions for the application of the rules of the calculus of probabilities, the more closely does the assumed error approximate to the actual error.

In any case, the probable or mean error gives a fairly good idea of the degree of accuracy attained. In the ordinary routine of daily observations, however, one is not usually concerned with the accuracy of the observations — this is a matter for consideration in the determination of geographic coordinates, parallaxes, masses of heavenly bodies, *etc.* Nevertheless, now that the radio-telegraph permits the transmission and reception of the exact time and consequently the systematic determination of differences of longitude, it is well to know the degree of accuracy of the time received and of the time determined on the spot.

On this subject I think there is some need for correcting the ideas held relative to the accuracy of the determination of local time. In 70 years much progress has been made in this direction, both in determining the time and in keeping it. It has been said that the accuracy attained in the transmission and reception of time by radio is notably superior to the accuracy of the determination of the time itself. Now, if that holds good at some places, it is not true for all observatories, especially for those which have good instruments, well installed, under a clear sky, and in charge of observers skilled in the manipulation of the so-called *impersonal micrometer*.

I have cited elsewhere[1] numerous examples of time determinations, made chiefly in connection with differences of longitude between various places. From these it is seen that since 1909 an accuracy

---

[1] Journ. Observateurs, 1928; Mem. Accad. Pontif. Sci., 1928.

of 0.01 seconds has been attained; at least the mean square error is of this order of magnitude and is often less.

The purpose of this article is to indicate the conditions favorable to the attainment of so high a degree of accuracy in time determinations and also in predicting the corrections to be applied to the readings of a good clock.

II

To determine time accurately by the transit of stars across the meridian, it is first of all necessary to lay aside large instruments (meridian circles or simple transit instruments). With such instruments it is almost impossible to reduce the azimuth constant to a very small value and especially to keep it so. The collimation cannot be eliminated by the reversal of the telescope, an operation which demands a certain amount of time and which can be applied only to circumpolar stars. Besides, when the telescope is reversed the azimuth changes. The level correction is not well determined, either by means of large levels or by the mercury bath.

It is necessary, therefore, to employ instruments with broken telescopes, which are not heavy and may be reversed in a few seconds. The diameter of the objective should be between 70 and 100 millimeters. A telescope of 95 millimeters aperture permits the observation with an illuminated field of stars of the 7th magnitude, and even of magnitude 7.5, during twilight.

It has been proposed recently to use straight telescopes with zenithal eye-piece; but, for one thing, this eye-piece has its inconveniences; for another, the straight telescopes are heavier and the diameter of the objective must therefore be reduced to 70 millimeters, or to 75 millimeters at most. Now, to have the determinations of the time close together, one must profit by all the periods of clear sky, observing sometimes during twilight. With telescopes of 70 millimeters aperture small stars may be observed with an illuminated field during the night only.

Someone may say that the catalogue of fundamental stars does not give the places of the stars down to the 7th magnitude; but I have already proposed elsewhere[1] that in order to confine the observations to stars rather close to the zenith, each observatory should make for itself a list of its own stars and carefully determine their right ascensions, referring them to the same system.

If the instrument chosen possesses good levels, enabling the air bubble to be adjusted to a nearly constant length, then by reading them with all the recommended precautions one may rely upon the value of the inclination. The level will be read for each star, and, of course, the observer will try to reduce the azimuth and the inclination to a minimum. The effect of the azimuth is almost null on the stars culminating within a few degrees of the zenith, not more than 25°. As to the impersonal micrometer, the observer must learn to use it perfectly, otherwise, as experience has shown, the personal equation is not eliminated. Moreover, beyond a declination of 60° the impersonal micrometer adds nothing to the accuracy obtainable by the ordinary micrometer.

Let us examine now the degree of accuracy that may be attained. In the clock correction determined by *one* star there remains:

1. The residual error of the apparent right ascension. It may be considered to amount at the most to ±0.°02, if the observer employs the star places of the Auwers' *New Fundamental Catalogue* corrected by A. Kopff, Director of the Rechen-Institut of Berlin.
2. The azimuth error, which for the zenithal stars amounts at the most to ±0.°006.
3. The error of the inclination, which is of the order of ±0.°01.
4. Finally, the error of the observation itself, which is a minimum, since with the impersonal micrometer the observations of different astronomers agree well. Let us assume ±0.°004 for this error.

The total error will be

$$\sqrt{0.\overline{02}^2 + 0.\overline{006}^2 + 0.\overline{010}^2 + 0.\overline{004}^2} = \pm0.°0235$$

But assume even ±0.°03. It follows that in observing 9 stars the error to be feared in the clock correction is only ±0.°01. Even if the observer did not have special skill, even if the atmospheric conditions were not completely favorable, etc., it will always be granted that except in unusual cases, with 15 or 16 stars an observer of moderate skill will determine the correction Cp or Δt for the clock with an error of ±0.°01.

Seventy years ago observations were made to the nearest second!

III

But astronomers are not satisfied with determining the exact time for a given instant. They must be able to give it at any instant. They must have a time-keeper. Today good Riefler clocks, kept at constant pressure and temperature, for several days following a direct determination give the time with an uncertainty of $0.°02$ or $0.°03$. But, as with the broken telescopes of the Bamberg type, one must know how to use these clocks and how to get from them all that is possible in the way of accuracy. It is known that by means of a pump one may change the air pressure in the metal case in which the pendulum swings.

To be able to interpolate and extrapolate the exact time when several determinations of the time are available, it is necessary to determine the clock-rate and to use first and second differences.

Abrupt variations, "jumps," in the rate of a good Riefler clock are improbable. In any case, it is sufficient to have another control clock beside the master clock, which enables abrupt variations in the rate of the latter to be detected. The next direct determination of the time will permit the elimination of any uncertainty.

IV

I submit here an example furnished by Riefler clock no. 60, the corrections for which have been furnished me by the Superintendent of the Naval Observatory at Washington. I here thank him for them.

These corrections come from the time determinations made with the small, straight, Prin telescope whenever the condition of the sky permitted. I believe that by using a Bamberg model broken telescope and the Kopff right ascensions, a greater accuracy would have been attained. The mean error would be only $\pm0.°01$. Furthermore, clock no. 60 is not of the most recent model.

Let us consider first one point. It is said that by determining the clock rate by two successive determinations of the time spaced 7 or 8 days apart, the residual errors of these determinations are reduced as if by dividing them by 7 or 8. For example, if the mean error of one determination of the time is $\pm0.°01$, the error of the rate during 10 days is only

$$\frac{0.°01\sqrt{2}}{10} = 0.°00141 \quad . \quad . \quad . \quad .$$

But this supposes that the clock rate has been constant during 10 days, which is not the case. The value obtained for the rate, supposing it to have a linear variation, applies exactly only at the epoch midway between the two dates corresponding to the two determinations of the time. It is the mean rate that is found; and when the question arises of giving the correction for the clock for an intermediate date—for example, 5 days after the first date—this correction, calculated from the mean rate, contains:

1. The error of the first determination of the time.
2. The difference between the *actual* rate during the 5 days following and the *mean* rate multiplied by 5.

In the same way, in predicting the correction for the clock for a later date, the mean rate during 10 days is not as exact as that which one would obtain with two determinations of the time spaced 3 or 4 days apart. The conclusion is that it is necessary to observe as often as possible and to determine the rate by means of first and second differences.

As to abrupt variations, they are more probable in an interval of 10 days than in one of 3 days.

The table which follows contains for a complete administrative year —July, 1926, to June, 1927—the daily rate of Riefler clock no. 60 of the Naval Observatory and the epochs to which they correspond. To construct this table from the series of clock-corrections that were so kindly supplied to me, I have grouped two by two all the successive corrections by taking the mean. I have likewise taken the means of the dates to which these values of Cp correspond. Then I have taken the first differences between these means of the Cp and I have done the same for the dates. Finally I have divided respectively the first, which are the variations of the Cp, by the second, which are the corresponding intervals of time. Since small intervals were concerned, I was entitled to assume that the values of Cp varied linearly. The rates given in the table thus apply to epochs midway between the two means of the dates that correspond to them.

An inspection of this table is very instructive. It shows that the rate of the clock varies very slowly and that the difference between two successive rates is, on the average, $\pm 0.°005$ or $\pm 0.°006$. These differences are due to the residual imperfections of the values of Cp and also to the small variations in the clock rate. Rarely does this difference reach $0.°015$. Only during the spring are there greater variations, which are doubtless due to the variability of weather conditions. Perhaps the temperature has not been kept absolutely constant, or the pressure has not been adjusted every time that the Cp showed the necessity for it. We may conclude that today, with two good instruments available, in charge of skillful astronomers, and well-installed, under a sky which permits the determination of the time, on the average, every 2 or 3 days, the value of Cp may be found with a mean error of

$$\frac{\pm 0.°01}{\sqrt{2}} = 0.°0071$$

With one good clock checked against another we can forecast Cp for 3 or 4 successive days with an uncertainty amounting hardly to $\pm 0.°02$.

It is a splendid triumph for astronomy!

TABLE 1.—DAILY RATES OF RIEFLER CLOCK No. 60, AT THE NAVAL OBSERVATORY IN WASHINGTON, IN 1926-1927

1926				1927			
Date	Rate	Date	Rate	Date	Rate	Date	Rate
July 6, 579	−0.ˢ037	Oct. 14, 819	−0.ˢ050	Jan. 3, 825	−0.ˢ078	Mar. 31, 354	−0.ˢ039
10, 156	42	17, 767	51	6, 616	88	Apr. 5, 732	42
13, 316	40	20, 259	46	8, 636	65	10, 848	37
17, 053	30	23, 001	50	10, 898	59	13, 901	45
21, 300	34	26, 002	48	13, 671	62	16, 417	49
25, 370	48	29, 054	48	16, 417	70	19, 352	38
28, 934	34	31, 856	56	19, 352	65	22, 512	33
Aug. 1, 409	28	Nov. 4, 060	48	22, 900	70	25, 340	36
4, 392	43	6, 020	55	25, 340	67	28, 111	24
6, 609	39	8, 333	56	28, 164	64	May 1, 940	14
8, 066	26	11, 377	43	31, 871	66	3, 625	33
9, 399	33	14, 810	44	Feb. 4, 341	73	5, 099	25
17, 096	34	17, 765	47	7, 581	58	8, 046	33
24, 539	40	20, 281	45	10, 088	57	12, 856	17
27, 032	36	22, 775	38	16, 386	54	16, 435	18
31, 538	38	25, 262	40	21, 406	46	24, 212	18
Sept. 6, 569	41	27, 313	46	24, 913	56	27, 399	24
11, 356	35	Dec. 1, 056	34	27, 126	34	June 1, 365	30
14, 568	38	3, 557	25	Mar. 1, 128	79	2, 356	28
17, 305	39	7, 039	78	3, 640	56	7, 153	09
20, 071	39	11, 011	73	5, 650	50	8, 374	28
23, 301	42	16, 021	80	9, 171	45	12, 648	08
27, 794	38	22, 064	86	10, 902	41	13, 696	−0.ˢ006
Oct. 2, 542	45	27, 076	89	12, 862	29	15, 743	−0.ˢ001
6, 027	32	29, 072	78	15, 743	82	19, 989	+0.ˢ015
8, 819	45	31, 068	78	19, 125	52	22, 917	−0.ˢ023
11, 620	−0.ˢ060		−0.ˢ070	22, 742	46	23, 989	−0.ˢ023
				25, 767	34	26, 380	−0.ˢ032
					35	29, 165	−0.ˢ024
					−0.ˢ045		

52

Introduction to

Concerning the Origin of Chemical Elements

George Gamow (1904 – 1968), was a Russian-born theoretical physicist and cosmologist. He discovered alpha decay via quantum tunneling and worked on radioactive decay of the atomic nucleus, star formation, stellar nucleosynthesis, Big Bang nucleosynthesis, the cosmic microwave background, and genetics. He produced an important cosmogony paper with his student Ralph Alpher, which was published as "The Origin of Chemical Elements" (*Physical Review*, April 1, 1948). This paper became known as the Alpher-Bethe-Gamow theory (although Bethe was not an author on the paper).

The following article appeared in 1942 in the Washington Academy of Sciences Journal, Volume 32, and pre-dates the cosmogony paper. It addresses early thoughts on stellar evolution. The first nuclear sequence in the article represents lithium burning which proceeds very fast and is especially important in determining the age of pre-main sequence stars. Stars, which achieve the high temperature (2.5×10^6 K) necessary for fusing hydrogen, rapidly deplete their lithium. This occurs by a collision of lithium-7 and a proton producing two helium-4 nuclei as described in the article. The temperature necessary for this reaction is just below the temperature necessary for hydrogen fusion. Convection in low-mass stars ensures that lithium in the whole volume of the star is depleted. Therefore, the presence of the lithium line in a candidate brown dwarf's spectrum is a strong indicator that it is indeed substellar. The CNO cycle (the second nuclear sequence in the article) is appropriate for heavyweight stars during their main sequence lifetime. Gamow then speculates on the issue of heavy element production.

A curve of growth (or abundance curve) expresses how the width of a spectral line (representing a particular element) changes with the number of atoms of that element.

Concerning the Origin of Chemical Elements

G. Gamow

The George Washington University

It is well known that the chemical analysis of the universe indicates a striking uniformity in the distribution of various chemical elements. In fact, we know that the meteorites, which most probably represent the fragments of some old broken-up planet, possess nearly the same proportions of various elements as the samples of terrestrial material, and that the spectral analysis of our sun and other stars leads again to a very similar chemical constitution.[1] It may be added that the recent investigations of interstellar absorptions indicate that approximately the same chemical constitution should be also ascribed to the extremely rarified gaseous material filling up the interstellar space.

Considering the known abundances of various elements from the point of view of possible nuclear transmutations, we should ask ourselves first of all whether these abundances are due to some nuclear processes taking place *at present* in various parts of the universe, or whether the abundance-curve should be considered as a "frozen-distribution" corresponding to some unusual conditions that existed in the *early creative stage* of the universe? The recent study of the problem of stellar energy sources shows quite definitely that *some features* of the abundance-curve are of more or less contemporary origin and can be understood on the basis of thermonuclear reactions taking place in the hot interior of stars. Thus, for example, we know that light elements lithium, beryllium, and boron are subject to rather rapid destructive reactions in the presence of hydrogen at the temperatures ranging from 5 to 15 million degrees. These thermonuclear reactions proceed according to the equations:

---

[1] The only large discrepancy between the chemical constitution of stellar and terrestrial material consists in comparatively large abundance of hydrogen and helium in stars (35 percent H and at least a few percent of He) as compared with the earth (0.001 percent of H, and 0.000,000,000,1 percent of He). There is, however, no doubt that this large discrepancy in abundances is of purely secondary character and is entirely due to the fact that H and He, being the light gases, had much better chance to escape from the terrestrial atmosphere into the surrounding empty space.

$$^6\text{Li} + {}^1\text{H} \rightarrow {}^4\text{He} + {}^3\text{He}$$
$$^7\text{Li} + {}^1\text{H} \rightarrow 2\,{}^4\text{He}$$
$$^9\text{Be} + {}^1\text{H} \rightarrow {}^6\text{Li} + {}^4\text{He}$$
$$^{10}\text{B} + {}^1\text{H} \rightarrow {}^{11}\text{C} + h\nu$$
$$^{11}\text{B} + {}^1\text{H} \rightarrow 3\,{}^4\text{He}$$

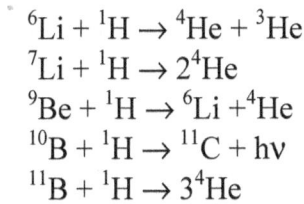

and result in the complete destruction of the elements in question and the formation of the large amounts of thermonuclearly inert helium. It was suggested by Gamow and Teller[2] that these particular reactions represent the main source of energy in the early stages of stellar evolution (in the so-called red-giant-stars), and that entering the main sequence the star must have these three elements completely destroyed in its interior regions. Although in the outer layers of the star the temperature is not high enough to induce such reactions, a certain amount of these elements must have been removed by diffusion from the stellar atmospheres, a fact that explains the anomalous drop in the corresponding region of the abundance-curve. In the next, main-sequence stage of stellar evolution the temperature in the interior rises up to 20 million degrees, inducing thermonuclear reactions of the next two elements, carbon and nitrogen, which, according to Bethe,[3] undergo the following transformations:

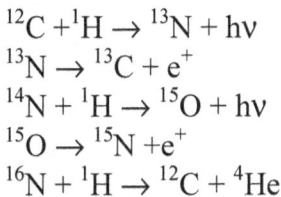

$$^{12}\text{C} + {}^1\text{H} \rightarrow {}^{13}\text{N} + h\nu$$
$$^{13}\text{N} \rightarrow {}^{13}\text{C} + e^+$$
$$^{14}\text{N} + {}^1\text{H} \rightarrow {}^{15}\text{O} + h\nu$$
$$^{15}\text{O} \rightarrow {}^{15}\text{N} + e^+$$
$$^{16}\text{N} + {}^1\text{H} \rightarrow {}^{12}\text{C} + {}^4\text{He}$$

We see from these formulae that carbon and nitrogen are not being completely destroyed by the reaction, but are constantly regenerated, thus serving only as some kind of catalysis for the transmutation of hydrogen into helium. The above reactions, however, serve to establish a definite balance between the relative abundances of ^{12}C — ^{13}C and ^{14}N — ^{15}N isotopes. For the temperature and pressure existing in stellar interiors, the equilibrium-proportions of these isotopes have been calculated by Bethe to be 70:1 and 500:1, which is in fair agreement with the observed abundances.

In spite of these successes in understanding the features of the

[2] G. GAMOW and E. TELLER. Phys. Rev. 55: 791. 1939.
[3] H. BETHE. Phys. Rev. 55: 434. 1939.

abundance-curve for lightest elements, however, the situation becomes much more difficult in the case of heavier elements. In fact, there seems to be no doubt that the much higher temperatures, needed for the transmutation of heavier elements, are not to be found in stellar interiors or, for that matter, in any other part of the present state of the universe.[4] Thus the only possible way to understand the origin of heavy, and particularly of radioactive elements lies in the assumption that in some previous stage of the development of our universe, physical conditions have been in general entirely different from what they are now, and that the temperature and density of matter then were, as a rule, much higher. Such a hypothesis finds strong support in the theory of expanding universe, according to which the matter, which is at present distributed rather rarely through space, expanded from the original state of very high density and temperature. It is particularly interesting that, according to the measured rate of present expansion, these extraordinary physical conditions in space must have been existing only about 2 or 3 billion years ago, a period of time comparable with the life-period of the long-living radioactive elements (thorium and uranium), which are still in existence.

Considering the present abundance of elements as the result of these long-past conditions, one can follow two different points of view: (1) That the abundance-curve corresponds to a thermodynamic equilibrium state at some very high density and temperature, which existed during a certain early expansion-stage of the universe; and (2) that relative abundances of various elements are due to a non-equilibrium breaking-up process of the original bulk of nuclear matter caused by a rapid expansion in the early evolutionary stages.

A detailed study of the first possibility was carried out recently by Chandrasekhar and Henrich,[5] who calculated the equilibrium-numbers of various nuclei corresponding to conditions of extremely high densities and temperatures. In these calculations, which extended over the first 20 elements of the periodic system, the authors took into account the exact values of the mass-defects of the nuclei in question with the interesting result that the theoretical abundance-curve repeats rather exactly all local

---

[4] An attempt to understand the building-up (Aufbau) process of heavy elements at the comparatively low temperatures existing in stars was made by Weizsaker (Phys. Zeitschr. 39: 633. 1938), but it turned out to be completely unsuccessful and has been entirely abandoned.

[5] S. CHANDRASEKHAR and L. R. HENRICH. Astr. Journ. 9: 288. 1942.

irregularities of the empirical curve. It must be noticed, however, that this particular result does not necessarily speak in favor of the thermodynamic-equilibrium hypothesis, since also in the case of a rapid breaking-up process more stable nuclei should have been produced in larger quantities than the less stables ones.

In respect to the general behavior of the abundance-curve, the results are considerably less satisfactory, which is owing exclusively to a very peculiar behavior of the empirical curve. In fact, the general habitus of this curve can be characterized as a *rapid exponential decrease up to the middle of the periodic system, and an approximate constancy in the second half of it.* (See figure.)

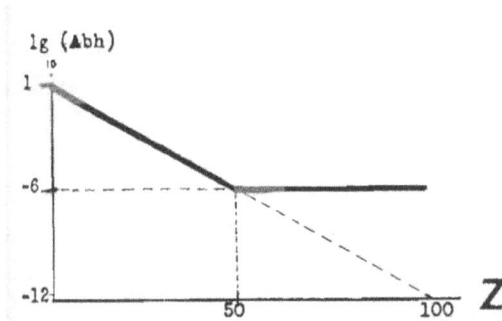

This character of the exponential curve excludes any possibility of understanding the abundance of all elements as the result of some kind of equilibrium, since in choosing the temperature and density so as to fit the decreasing half of the curve (Chandrasekhar and Henrich assume $\rho = 10^7$ gm/cm^3 and $\tau = 8.10^9$ °C), one should necessarily expect the continuation of such decrease also for the group of heavier elements. There also seems to be no physical possibility of explaining the peculiarity of the empirical curve by any kind of "freezing up" of heavier elements while the lighter ones are still being transformed. In fact, any such transformation should be necessarily connected with the emission of a large number of high-energy neutrons, which are bound to affect the relative number of heavier elements, and to cause the later part of the curve to drop down.

We can try now to investigate the second possibility, and to see which kind of distribution could be obtained from the hypothesis of a rapid breaking up of the original superdense nuclear matter. We must remember that, according to our present knowledge of nuclear fission-processes, all nuclei that are heavier than uranium would be immediately broken into two or more approximately equal parts (slight deviation from equality of fission-fragments being due to the irregular character of the mass-defect curve). If we assume for a moment that each unstable superheavy nucleus breaks up in only two approximately equal parts, the statistical result of such a breaking up process, will evidently correspond to *an equal abundance of all elements belonging to the second half of the periodic system, and to a complete absence of all lighter elements.*

This gloomy picture may be improved, however, if we remember that: (1) Even in ordinary uranium-fission *a number of free neutrons are being emitted in each breaking-up process, and this number most probably increases in the case of the more violent fission of superheavy nuclei.* Neutrons produced this way will turn spontaneously into protons, and will contribute to a larger abundance of hydrogen. (2) Although (in the known fission-processes of radioactive elements), the nucleus always breaks in only two fragments, *we may expect that for the heavier nuclei the probability of triple and higher order splitting is considerably larger.* Such multiple splitting of nuclei several times heavier than uranium will not much affect the equipartition in the second half of the periodic system, but will, on the other hand, produce a large number of lighter nuclei.

It seems, therefore, on the basis of the foregoing remarks, that this possibility is not entirely excluded, that the main features of the abundance curve may be explained as the result of a complicated fission-process. In order to answer this question in a more definite way it will be necessary to study the stability of superheavy nuclei with respect to multiple-fission and to investigate the statistical distribution of the fission-fragments in a successive breaking down process. The work in this direction is now in progress, and its results (if any) will be published later.

Introduction to

Cepheid Variables and the Period-Luminosity Relation

Cecelia Payne-Gaposchkin (1900 – 1979) was an astronomer whose PhD dissertation was said to be the best of the twentieth century. She was the first to show that the Sun, like all stars, is composed mainly of hydrogen. The following article appeared in 1959 in the Washington Academy of Sciences Journal Volume 49.

Today Cepheid variable stars are one of the important rungs on the cosmic distance ladder (see Washington Academy of Sciences Journal, Vol. 97, 2011 for a discussion of the cosmic distance ladder). The Cepheid class of stars is named for the prototype star δ Cephei. Cepheids vary in absolute luminosity with a predictable regularity. Their luminosity closely correlates with their period of variability (period-luminosity relation). Therefore, once the period of variability of a particular Cepheid is observed, its absolute luminosity is known. One also observes the apparent brightness, m, of the Cepheid. Brightness follows an inverse square law. One now has enough information to obtain the distance to that Cepheid. From the absolute luminosity, M, and the observed apparent magnitude (m), the distance (r) to that Cepheid is determined using the distance modulus $m - M = 5\log_{10}r - 5$. Hence the importance of Cepheids in setting celestial distances. Henrietta Leavitt was the first to discover the correlation between Cepheid luminosity and variability (in 1912). In the following article Cecelia Payne-Gaposchkin goes into detail about the difficulties inherent in establishing the correlation. The RR Lyrae class of stars she mentions is also used for distance calibration but is not as reliable as Cepheids. Today most of the uncertainties tied to the Cepheid distance scale are: the nature of the period-luminosity relation in various passbands, the impact of metallicity on both the zero-point and slope of those relations, and the effects of photometric contamination (blending) and a changing (typically unknown) extinction law on Cepheid distances.

Cepheid Variables and the Period-Luminosity Relation[1]

C. PAYNE-GAPOSCHKIN

Harvard College Observatory
(Communicated by Charles M. Herzfeld)

The period-luminosity relation has almost reached its fiftieth anniversary. Today it is being studied more actively than ever, and its observed complexities and theoretical implications are still far from exhausted. In fact I hope to show that what we know and understand at present about the period-luminosity relation is far less than what remains to be discovered and interpreted.

Introduction — The diagram derived by Miss Leavitt (1) from 25 Magellanic Cepheids, reproduced in Fig. 1, already shows several of the features of current interest: the linear relationship between the apparent magnitude and the logarithm of the period, the scatter of the points about the curve, and the variety of amplitudes. Each, as I shall describe, has been verified and amplified by later work.

The earlier work of Bailey (2) on the variable stars in globular clusters laid the foundation for the recognition of an equally striking relationship between period and apparent brightness within any one globular cluster. Although the variables with periods shorter than a day showed no marked correlation between period and brightness, the few with longer periods were always brighter. On the simple assumption that all the short-period variables were of similar luminosity, Shapley (3) constructed a period-luminosity curve from the data for all globular clusters that contained variables with periods longer than a day, and concluded that it had the same slope as the relation for the Magellanic Cepheids.

If the identity of the two period-luminosity relationships is granted, the conversion from apparent to absolute magnitude can be effected by the establishment of absolute magnitude for some one contributor to the curve. The short-period variables in globular

---

[1] The 28th Joseph Henry Lecture of the Philosophical Society of Washington, delivered before the Society on May 22, 1959.

clusters were identified with the RR Lyrae stars of the galactic field, many of which have periods and light curves exactly like those of the cluster variables. Accordingly the absolute magnitudes of the RR Lyrae stars were intensively studied, and the generally accepted value 0.0 was used to fix the zero point of the period-absolute magnitude relation.

The immediate application of this conclusion by Shapley led to his epoch-making study of the dimensions of the galaxy. Later revisions of this work have resulted from improvements in the scale of apparent magnitudes and from corrections for obscuration, not from revisions in the zero point of the absolute magnitudes.

The upper part of the period-luminosity curve was used to derive the distances of the Magellanic Clouds, and when Hubble (4) discovered Cepheids in NGC 6822, Messier 33, and Messier 31, the distances of these galaxies were similarly derived. These applications of the period-luminosity curve depended on the original working assumption that the relationships among the stars in globular clusters and those in the Magellanic Clouds and other galaxies were not only parallel but identical. That this assumption is not justified was shown by Baade (5) in his study of the Andromeda galaxy. The zero point of the period-luminosity curve for classical Cepheids was thus revised upward by about 1.5 magnitudes, and all distances derived by the use of the upper part of the curve were accordingly also revised upward.

Within any one globular cluster there is very little correlation between the apparent magnitude and period of the RR Lyrae stars, but the photographic (though not the photovisual) magnitude has been shown for example by Roberts and Sandage (6) to be slightly brighter for the variables of shortest period in Messier 3. That the absolute magnitudes of RR Lyrae stars may differ slightly in different globular clusters is suggested by Sandage (7) and the possibility that the accepted absolute magnitude of RR Lyrae stars may require downward revision is discussed by Arp (8).

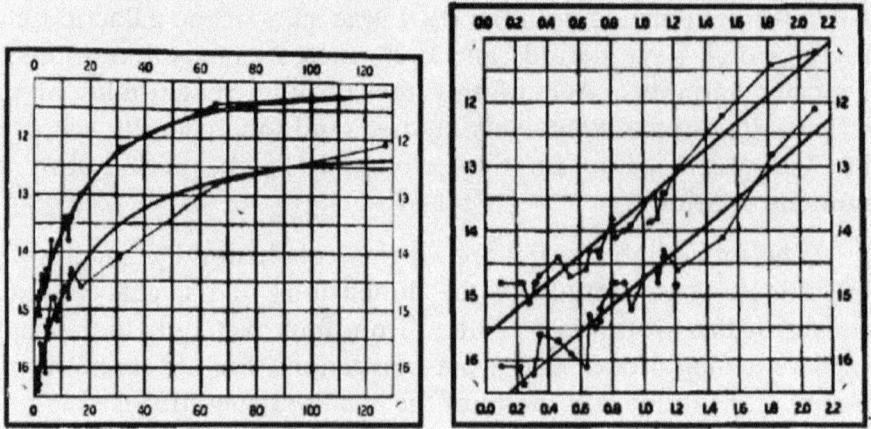

FIG. 1.—Miss Leavitt's period-luminosity relation (maximum photographic magnitude above, minimum below). Left: period plotted against magnitude; right, logarithm of period plotted against magnitude. From Harvard Circular 173. 1912.

The period-luminosity relations for variables in globular clusters and for Cepheid variables have been convincing separated by the work that has been briefly sketched in the preceding paragraphs. Another equally striking correlation that might well have pointed in the same direction many years ago is the Hertzsprung relationship between the periods and the forms of the light curves (9). The classical Cepheids of the galaxy show a progression in the form of a light curve (10), such that stars of the shortest periods show a smooth rise and fall, at longer periods (about 5 days) a hump appears on the declining side, and becomes more pronounced up to periods between 8 and 9 days. Between 9 and 10 days the curves abruptly become more symmetrical, with a more or less well-marked hump between two shoulders, and generally small amplitude. At rather longer periods the light curve has a more abrupt and asymmetrical rise, with a small hump just preceding the main brightening, and at the longest periods the curve is again found to be smooth and uncomplicated. The light curve thus runs through one complete cycle of changes between the shortest and longest periods. Cepheids in the Magellanic Clouds also display the Hertzsprung relationship (11).

An analogous relationship between period and light curve is shown by the short-period variable stars in globular clusters: in any one cluster, the stars of shortest period have light curves that are almost sine curves, and there is a rather abrupt transition at some longer period (not the same in all clusters) to a highly asymmetrical light curve, usually with a small hump before the main rise. Stars of the longest periods (less than a day) have less asymmetrical light curves. This relationship was pointed out by Bailey, who designated the curves by the letters *c, a, b,* in order of increasing period.

The Hertzsprung relationship for galactic classical Cepheids and RR Lyrae stars is illustrated in Fig. 2.

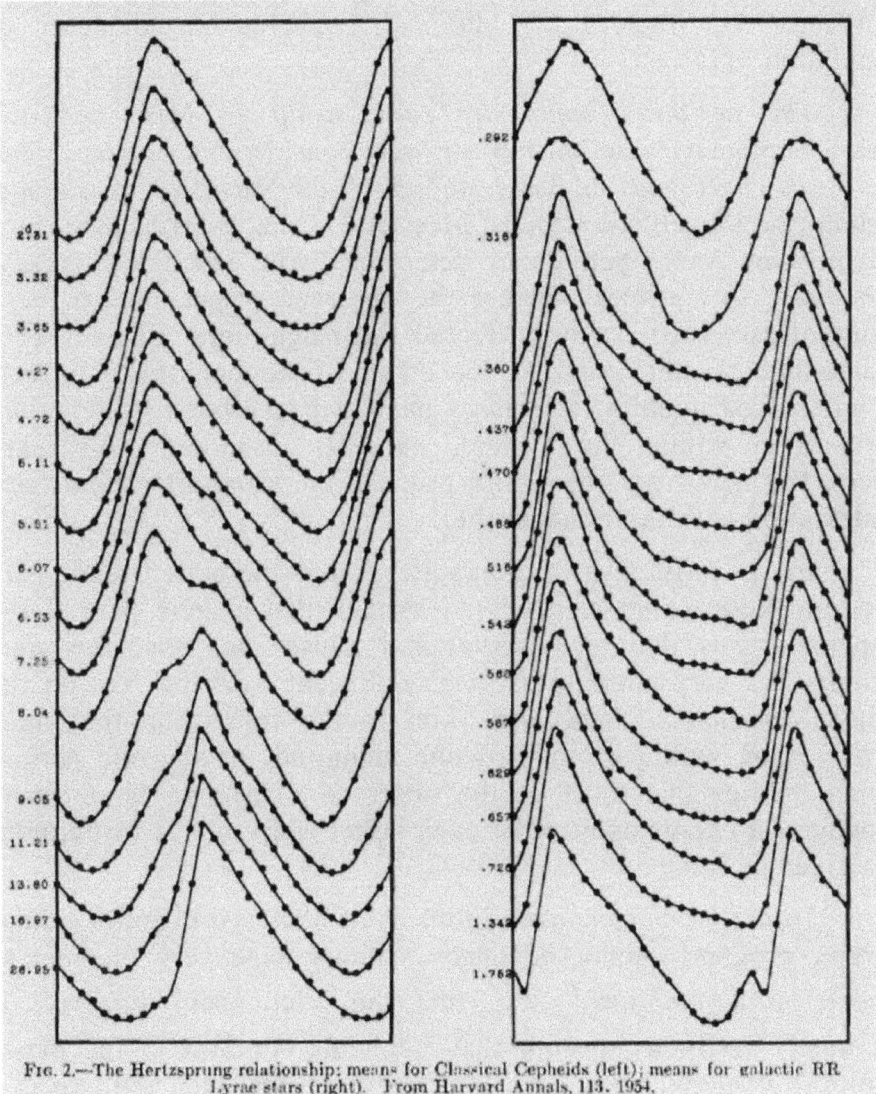

Fɪɢ. 2.—The Hertzsprung relationship: means for Classical Cepheids (left); means for galactic RR Lyrae stars (right). From Harvard Annals, 113, 1954.

The globular cluster stars with periods over a day do not conform to the Hertzsprung relationship; those with periods between 10 and 20 days show a broad maximum or a hump on the downward slope (12). A few galactic variable stars of similar period, which differ from classical Cepheids in lying far from the galactic plane, in motion, and in other ways, show similar light curves. In particular, a group of Cepheids associated with the galactic center shows a period-light curve pattern similar to that of the stars in globular clusters (13). These galactic variable stars are clearly to be associated, like those in globular clusters (14), with Population II, whereas the classical Cepheids are members of Population I (15).

The period - luminosity relationship — Miss Leavitt's period-luminosity relationship showed considerable scatter about the mean curve, and all later studies of the variables in a single system (the Magellanic Clouds, Messier 31, and so forth) also show a dispersion. As the periods are determined with adequate accuracy, it is usual to consider whether the observed dispersion can be a result of factors that have affected the magnitudes. Shapley (16) enumerated possible contributors: errors of the magnitudes, effects of unresolved doubles, effects of general background brightening, obscuration within the system, galactic obscuration, Eberhard effect. Erroneous periods would produce the same effect, but their contribution is certainly negligible.

In the early days of the application of the period-luminosity relation, there was a tendency to assume that a great part of the dispersion was due to observational causes and that the stars actually lay very close to the curve. Recent work by Arp on the Small Magellanic Cloud leads him to the conclusion that most Cepheids lie within a range of one magnitude at any one period, which implies that most of the observed dispersion is intrinsic. Sandage (18) contemplates the even larger range of 1.2 magnitudes at a given period.

Hitherto I have spoken in terms of observed quantities only. Further progress cannot be made without some very elementary theoretical assumptions: first, that the relationship $P\sqrt{p} = C$ is accurately fulfilled; second, that Cepheids conform to the mass-luminosity relationship. Granted these premises, then if there is a dispersion in magnitude at a given period, three conclusions follow:

1. At a given period, there is:
   a) a dispersion in color, the bluest stars being of highest luminosity
   b) a dispersion in luminosity, the brightest; stars being the bluest

2. At a given luminosity, there is:
   a) a dispersion in period, stars of longest period being the reddest
   b) a dispersion in color, the reddest stars being of longest period

3. At a given color, there is:
   a) a dispersion in period, stars of longest period being brightest
   b) a dispersion in luminosity, the brightest stars being of longest period

These three conclusions are shown graphically in Figs. 3, 4, and 5, which have been drawn for a range of magnitude of 1.0 at a given period, and for a range of color of 0.2 at a given magnitude.

The relation between mean color and absolute magnitude is not implicit in the assumptions, but can be shown to be plausible. The existence of a mean period-spectrum relation for Cepheids has long been recognized, and the data given by Code (19) define it accurately. If all Cepheids were of the same color, they would all have the same surface temperature. But the more luminous Cepheids would then have somewhat earlier spectral classes on account of lower surface gravity. Since the opposite is observed, we can infer that bright Cepheids are somewhat redder than fainter Cepheids and that the difference of color between two Cepheids is greater than the difference found between two stars of comparable luminosities and identical spectral class.

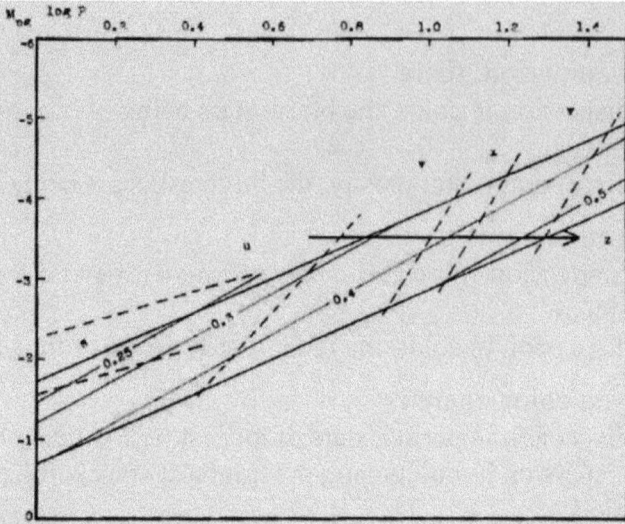

Fig. 3.—Relation between period and absolute magnitude (schematic); light lines show equal colors; broken lines separate the domains of different curve types. The arrow shows the possible course of Cepheid development.

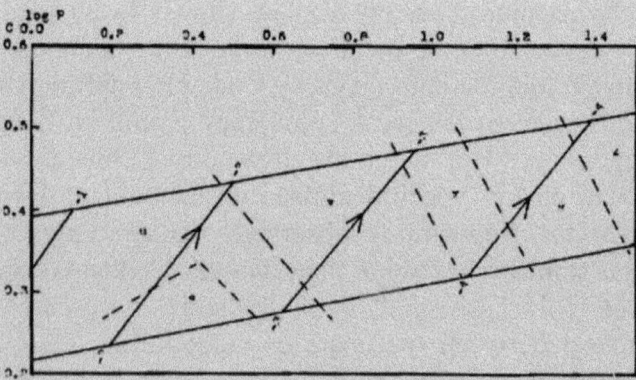

Fig. 4.—Relation between period and color (schematic). Light lines show equal absolute magnitudes; broken lines separate the domains of the different curve types. Arrows show the possible course of Cepheid development.

66

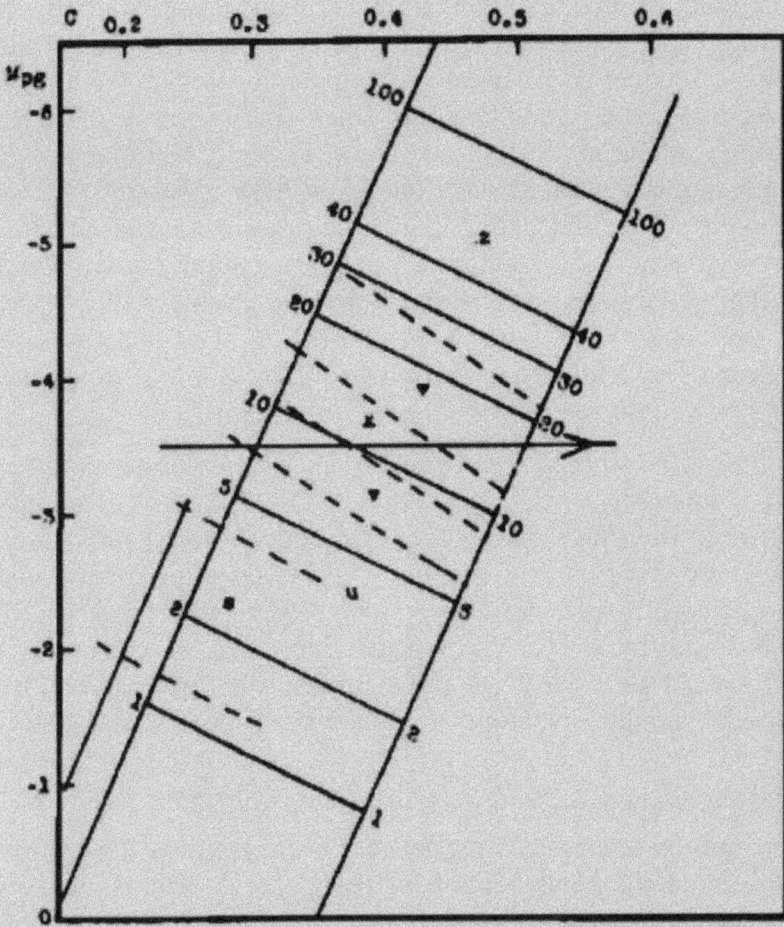

FIG. 5.—Relation between color and absolute magnitude (schematic). Light lines show equal periods; broken lines separate the domains of the different curve types. The arrow shows the possible course of Cepheid development.

Data on the accurate colors of Cepheids in other galaxies are needed, both to test the qualitative statements just made and to provide quantitative material from which consistent period-magnitude-color arrays may be constructed. However, the effect of obscuration also enters into such colors, and even if all Cepheids in a system were of the same true color at a given period, obscuration and extinction would conspire to make the most reddened stars the faintest. The size of the effect in our own system may be judged from the diagram of maximal color index against logarithm of period reproduced by Walraven, Muller and Osterhoff (20). The diagram contains a few points for the two Magellanic Clouds, from measures by Gascoigne; these, and the few colors published or discussed by Gascoigne and Kron. (21), Gascoigne (22), and Gascoigne and Eggen (23) are not inconsistent with the view here expressed. Gascoigne and Eggen act on the belief that the evidence for identity in color between galactic and Magellanic Cepheids, though not conclusive, is encouraging. Opinion has long been divided on this matter, and I will state my own view—that the colors of the galactic and the Magellanic Cepheids are essentially similar, and that the dispersion of color is as described above in simplified terms.

The Hertzsprung relationship — The relation between the period, luminosity, and light curve of a Cepheid must next be considered. The fact that averaged light curves for several stars with a small range of period display the Hertzsprung relation shows clearly that form of light curve is closely related to period. If the period-luminosity curve has a dispersion, then, if stars of a given period should have identical light curves, stars of a given absolute magnitude should have light curves that differ systematically with period.

The Cepheids in the Small Magellanic Cloud furnish information on this point. In order to present the data something must be said about the classification of light curves. The Bailey types *a* and *c* for the variables in globular clusters are excellent criteria, unambiguous and mutually exclusive. Type *b*, however, grades into type *a* and today is usually combined with it. In the classification of light curves, as in other matters (period frequencies, absolute luminosities of RR Lyrae stars) a globular cluster appears, so to speak, to lack a dimension in comparison to a system like the Small Cloud. This dimension is very likely mass: stars on the horizontal branch of a globular cluster must all be of almost the same mass. The Cepheids in a large system also differ in the time dimension. Methods of analysis and classification that give clear-cut results for globular clusters will not necessarily do so when applied to more complex groups. This statement, obvious in regard to color-magnitude arrays, is no less true of the classification of light curves. Methods that suffice for RR Lyrae stars are not flexible enough for Cepheids.

The first photoelectric light curves of Cepheids confirmed the variety and complexity of the light curves and color changes. Eggen (24) first defined groups A, B, C to represent the light curves, on the basis of relationships between period and amplitudes of light and color. Whether or not it was intended, the names suggested a parallel with the Bailey types, which has no physical justification. In a later paper, the definitions for types A, B, and C have been modified: a star is of type C if the light curve is symmetrical ($M - m = 0.37\ P$), of type B if a hump is present, of type A if no hump is observed (25).

To me it has always seemed that the Hertzsprung relationship is fundamental in describing the light curves of Cepheids, and that three classes are not enough to cover its complexities. My own classes have been defined as follows (26)

u: smooth, asymmetrical (δ Cephei)
u: smooth rise, hump on decline (η Aquilae)
w: saddle-shaped curve (SX Velorum)
x: central peak, shoulder on each side (Z Lacertae)
y: sharp rise, preceded by small rise (SZ Aquilae)
z: smooth, asymmetrical U Carinae)
s: sine curve (GH Carinae)

Eggen's type C covers types s and x, his B covers v and w, and his A covers u, y, and z.

The period-luminosity curve for minimal photographic brightness in the Small Magellanic Cloud is shown in Fig. 6, which embodies a great deal of unpublished material kindly made available by Dr. Shapley, for over 600 stars. The magnitudes are on the uncorrected Harvard scale; Arp's photoelectric work has shown that the range of magnitudes should be increased.

The points corresponding to stars with different curve types have been plotted with distinctive symbols, and I have indicated by broken lines the approximate domains occupied by light curves of different types.

An earlier diagram of the same kind (26) made with about a hundred stars from both clouds shows the effect more clearly. This is in part because the light curves were better determined, in part because the stars were selected for freedom from obvious obscuration. The reality of the dependence of curve type on luminosity as well as on period was verified by means of the χ^2 test.

Points for curve type *u* could not be included in the discussion on account of incompleteness at faint magnitudes.

A feature of Fig. 6 that does not run parallel to the other curve types is the group of symmetrical curves for stars of short period. These small-range curves seem to follow a period-luminosity spread parallel to the main one, and overlapping it slightly. These stars are of such interest that they must be examined critically to see whether their small range, and comparatively high luminosity for their period, could result from the presence of unresolved companions. They seem to be too numerous for this to be a likely interpretation. Moreover, they have counterparts in our galaxy (GH Carinae, FF Aquilae, DT Cygni).

Cepheids of similar curve type, therefore, are not aligned precisely with Cepheids of similar period. In Figs. 3, 4, and 5, the domains of similar curve type are separated by broken lines. These domains cut lines of equal period (mean density), equal luminosity, and equal color, at various angles, so curve type is not primarily determined by any of these properties of the star.

Development of RR Lyrae stars — The place of the RR Lyrae stars in stellar development has been deduced from their well-defined position in the H-R diagrams of globular clusters. They are evidently members of the horizontal branch; the continuity of star counts through the variable star domain, in variable-rich clusters, suggests that they move along the horizontal branch (27). The correlation of period and color found in Messier 3 by Roberts and Sandage implies that an RR Lyrae star changes its period as it crosses the gap, and makes an abrupt transition to or from "overtone" pulsation as it crosses the interface between the type *a* and *c* curves. Changes of period are not (and are unlikely to be) large enough to indicate which way the RR Lyrae star travels in the H-R plane although one can infer that many writers think in terms of progress from long to short period, from fundamental to overtone, from right to left in the HR plane.

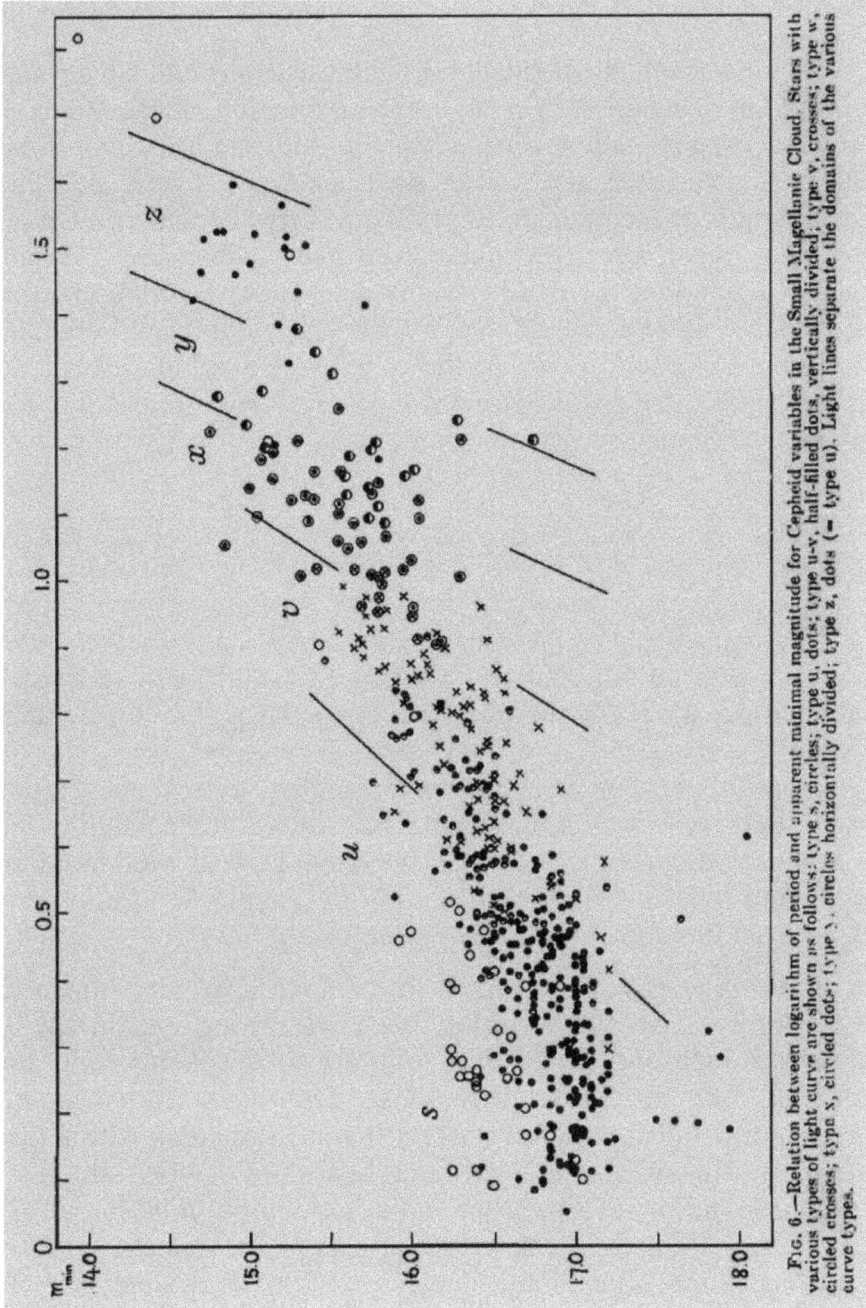

FIG. 6.—Relation between logarithm of period and apparent minimal magnitude for Cepheid variables in the Small Magellanic Cloud. Stars with various types of light curve are shown as follows: type s, circles; type u, dots; type u-v, half-filled dots, vertically divided; type w, circled crosses; type x, circled dots; type y, circles horizontally divided; type z, dots (= type u). Light lines separate the domains of the various curve types.

71

If the method of mapping empirical evolutionary tracks employed by Sandage (28) is valid, the distribution of stars along the horizontal branch might give a clue to the rate at which a star crosses the variable gap, or at least the relative rate at which different parts of the path are traversed. As Arp (29) has pointed out, a strongly populated horizontal branch goes with a large variable star population, but there are variable-poor clusters such as Messier 2 in which only one side of the variable star gap is well populated. Within the gap itself, however, the variable stars are not distributed uniformly with period. In ω Centauri, for example, numbers of stars within equal limits of log P are as follows:

log P	0.4 to 0.5	0.5 to 0.6	0.6 to 0.7	0.7 to 0.8	0.8 to 0.9	0.9 to 1.0
Number of stars	6	31	15	38	25	10

These numbers do not change uniformly, as might be expected for steady progress across the gap; rather they suggest acceleration-deceleration-acceleration as the gap is crossed. For classical Cepheids the situation is even more complex. In any one globular cluster there is a very small dispersion of absolute magnitude at any one period, but in a large stellar system such as the Magellanic Clouds and Messier 31, the dispersion in magnitude at one period adds at least one dimension to the problem.

Development of Cepheids — It has long been recognized that the Cepheid variables lie within the Hertzsprung gap in the HR plane. Although the true colors of galactic Cepheids, and their dispersion, are extremely difficult to determine, as has already been mentioned, it seems very likely that all Cepheids with a given luminosity lie within a restricted range of color (about 0.2 magnitude), and that no other stars lie within this range. The Population II Cepheids seem to occupy a similar domain, probably somewhat to the blue of the domain for classical Cepheids. The situation as I see it is shown in Fig. 7. The variable supergiants appear to lie on either side of the Cepheid domain (30).

The picture has been refined by the discovery of a few Cepheid variables in galactic clusters (31). The well-established members are tabulated below.

72

Cluster	Cepheid	Period	Light Curve	
			Eggen	This paper
NGC 6664	EV Sct	3.09	C	s
NGC 7788	CE Cas a (32)	4.45		
	CF Cas	4.87	AB	u
	CE Cas b	5.13		
Messier 25	U Sgr	6.74	AB	v
NGC 129	DL Cas	8.00	C?	s?
NGC 6087	S Nor	9.75	C	x

The suggestion of Bidelman that UX Persei ($4^d.57$), UY Persei ($5^d.37$), VX Persei ($10^d.90$) and SZ Cassiopeiae ($13^d.60$) are members of the Perseus I association should be mentioned (33). The Burbidges regard SZ Cassiopeiae as "probably a member of the cluster." The apparent magnitudes of all four stars are correlated with their periods.

The stars in the table are all as bright as, or brighter than, the brightest main sequence stars in the clusters (34); the four possible members of the Perseus cluster are fainter than the bright B stars and M stars in the Perseus double cluster. These data suggest that the age of a Cepheid in a galactic cluster can be determined:

Star	Cluster	Age of cluster years	Age of Association years	Ref.
EV Sct	NGC 6664	2×10^8		(33)
CE Cas a		10^8		(33)
CF Cas	NGC 7788	10^8		(33)
CE Cas b		10^8		(33)
U Sgr	Messier 25	10^8		(33)
DL Cas	NGC 129	10^8		(33)
S Nor	NGC 6087	10^8		(33)
SZ Cas	h, χ Persei	10^6		(35)
	Perseus I		4×10^6	(36)

The Cepheids in galactic clusters evidently lie within the Hertzsprung gap, and it is rather clearly indicated that a star becomes a Cepheid as it reaches a certain critical color as it crosses the gap. Analogy with globular clusters, where nonvariable stars are found on both side of the gap, might suggest that it emerges on the other side of the gap and becomes a red giant, and at least in NGC 6664 there is clear evidence of red giants (31). If SZ Cassiopeiae is a member of the Perseus Association, it again is associated with red giants, and ER Carinae has been suggested as a possible (though unverified) member of NGC 3532, which is very rich in red giants. We note also that the lower limit of the classical Cepheids is not far from the place (between NGC 752 and Messier 67) where the Hertzsprung gap has narrowed and disappeared. The general picture is shown in Fig. 8, which amplifies Sandage's well-known diagram by the addition of material for the Orion I association (37) and the Magellanic Clouds (38).

The three Cepheids in NGC 7788 display a rather small range of periods and magnitudes. The bright "galactic" cluster NGC 1866 in the Large Magellanic Cloud is an example of a cluster that contains many Cepheids, most of them in a restricted range of period (39). The following table gives the names, periods, and approximate mean magnitudes; stars within 10' of the center of the cluster are marked with asterisks.

HV	Period d	Mean m_{pg}		HV	Period d	Mean m_{pg}
12206*	2.506	16.1		12194*	3.205	16.0
12208	2.604	16.2		12205*	3.210	15.9
12199*	2.639	16.51		12189	3.246	16.4
12200*	2.725	16.4		12187	3.287	15.8
12209	2.930	16.21		12204*	3.439	15.6
12188	2.934	16.31		12201*	3.444	15.8
12211	2.940	16.21		12193	3.465	16.3
12203*	2.954	16.4		12198*	3.523	16.2
12202*	3.101	16.2		12207	4.566	15.8
12196*	3.113	16.4		12211	5.083	15.9
12197*	3.144	16.01		12186	12.24	15.0
12195	3.190					

Most of the Cepheids in NGC 1866 lie within the limits 0.40 and 0.55 in log P, and there is a trend toward brighter magnitudes for longer periods. The picture is like that presented by nearby galactic clusters, except that NGC 1866 is much richer in Cepheids. Details of its color-magnitude array will be of extreme interest. If there is a detailed analogy with galactic clusters, we can infer an age comparable with that of EV Scuti from the average period of the Cepheids. With the anticipated discovery of more Cepheids in Magellanic clusters, we can look forward to verification and extension of the dating procedure.

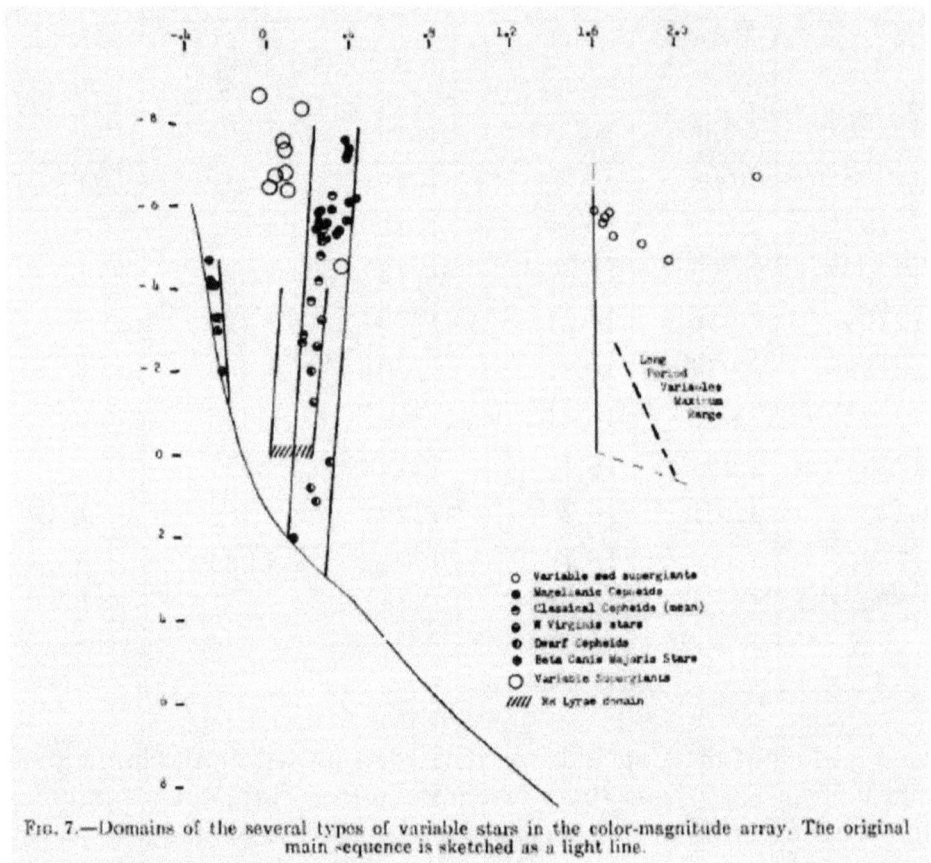

Fig. 7.—Domains of the several types of variable stars in the color-magnitude array. The original main sequence is sketched as a light line.

Galactic clusters provide suggestive qualitative information about the course of development of a Cepheid; among other things they suggest that the Cepheids follow a route like that of stars of mass over 2.5 suns, and move more or less horizontally across the HR plane, rather than rising sharply after leaving the main sequence, like stars of solar mass and less. This is the justification for the earlier assumption that Cepheids conform to the mass-luminosity relation.

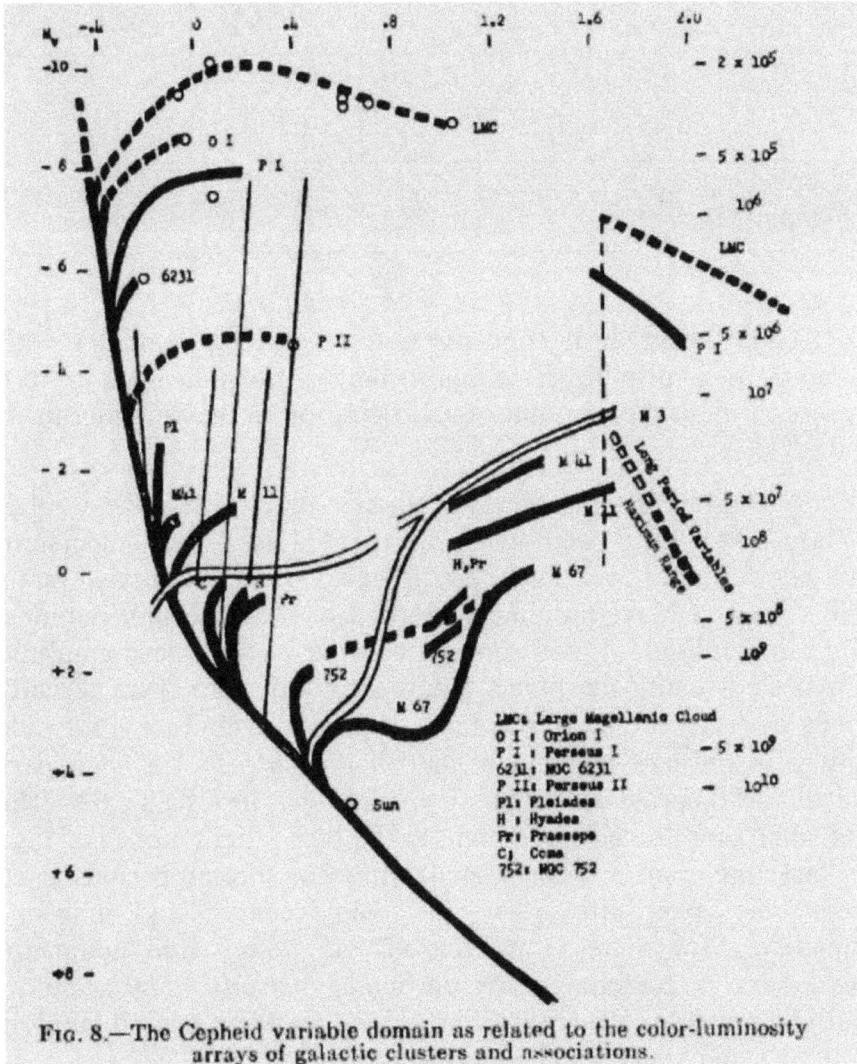

Fig. 8.—The Cepheid variable domain as related to the color-luminosity arrays of galactic clusters and associations.

In a globular cluster, there is a strong correlation between the amplitudes and periods of the RR Lyrae stars, which we have mentioned as probably traversing the horizontal branch across the variable gap: as we go from long to shorter period, the amplitude rises to a maximum, and then falls abruptly at the transition between curves of type a and type c. In NGC 1866 we note that four stars of amplitude 1.2 magnitudes and more lie between $\log P = 0.5$ and 0.54, but there are also two stars of large amplitude with $\log P = 0.41$, one of them within 10′ of the center of the cluster. The numbers of stars within equal limits of $\log P$ are:

Log P	0.4 to 0.45	0.45 to 0.5	0.5 to 0.55	0.65 to 0.7	0.7 to 0.75	Over 1.0
Number of stars	4	7	9	1	1	1

Thus the greatest concentration of periods coincides with a maximum of amplitude. Here again there is no indication of steady progress through the gap, but of a slowing-up in the neighborhood of log P = 0.525.

A single cluster presents a simpler picture than the Cepheids in a larger system; its stars may be regarded as having dispersion in mass, but do not differ appreciably in age. It is not surprising that the Small Cloud displays no simple relation between period, luminosity and mean amplitude. Fig. 9 shows a contour map of mean amplitudes in the period-luminosity plane. Small amplitudes are characteristic of periods between 8 and 10 days and apparent magnitude 15.4 (about absolute magnitude −3.6), and also of the strip to the short-period edge of the diagram for apparent magnitudes less than 16.0. These latter small amplitudes correspond to the type s light curves of Fig. 6. The fact that they are associated with the shorter periods speaks against the possibility that they are caused by unresolved companions, for there is no reason why unresolved companions should have a preference for particular *periods*; they should be equally common at all periods for a given apparent magnitude. Large amplitudes show an even more striking distribution. They occur, as is well known for the more luminous stars, but they also occur at the low-luminosity, long-period edge of the period-luminosity array. The occurrence of large amplitudes among the fainter Magellanic Cepheids has been noted by Arp and is undoubtedly a real phenomenon.

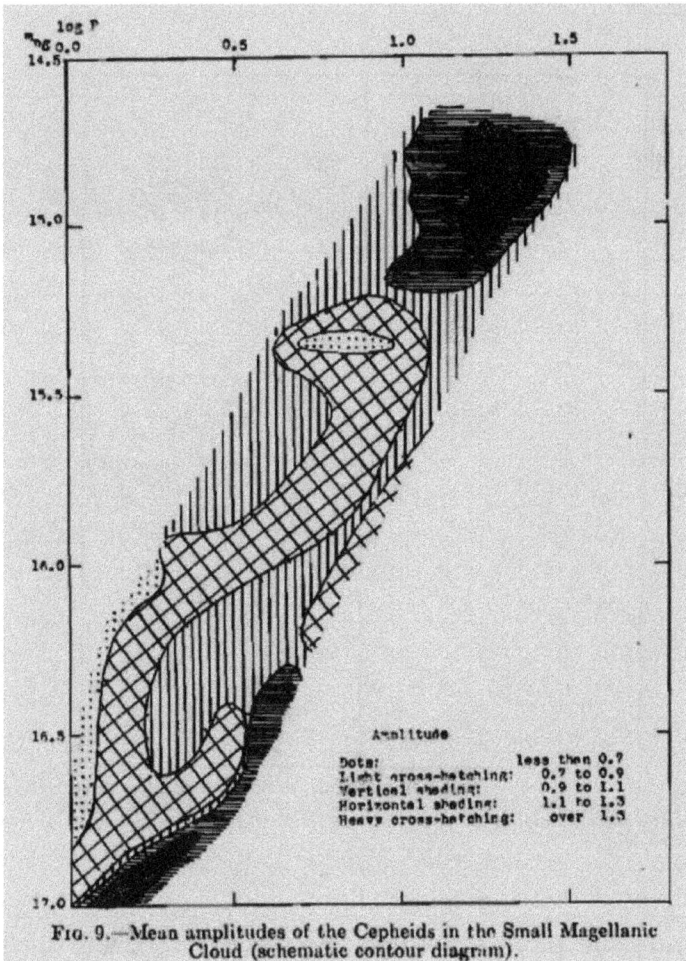

FIG. 9.—Mean amplitudes of the Cepheids in the Small Magellanic Cloud (schematic contour diagram).

If the path of any one Cepheid is a horizontal track across the period-luminosity plane, we might be able to trace the changes that a star undergoes by examining the changes in light curve at any one apparent (and therefore absolute) magnitude with changing period. Figs. 10, 11, and 12 show such series of light curves at seven different apparent magnitudes, and are typical for the whole available material. At brighter magnitudes, the light curve shows systematic changes and shifts of one prevalent pattern, which recall the Hertzsprung relationship, and illustrate the angle at which light curves of different types cross the average period-luminosity relation. For fainter magnitudes, however, the stars of shortest period tend to have small-amplitude, symmetrical light curves (type *s*), which recall the type *c* light curves for RR Lyrae stars in globular clusters.

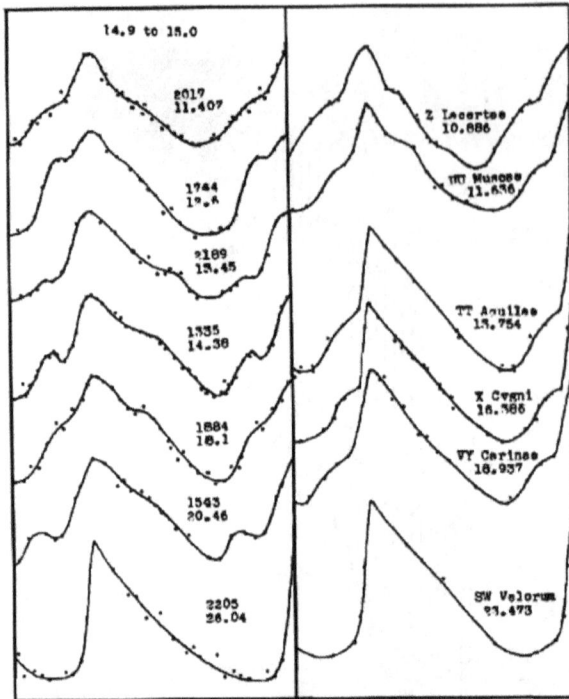

Fɪɢ. 10.—Light curves of Cepheids in the Small Magellanic Cloud, mean apparent photographic magnitude 14.9. Harvard Variable number and period are indicated for each. On the right, for comparison, several galactic Cepheids (photoelectric light curves by Eggen, except, for UU Muscae, Harvard photographic light curve).

If it is accepted that a Cepheid travels from left to right in the HR plane, and if the type *s* light curves represent overtone pulsations (as Sandage has suggested for the very similar Eggen type *c* curves), must we suppose that the developing Cepheid proceeds in the direction overtone to fundamental, short to long period?

With this possibility in mind we can reexamine the amplitudes of the Magellanic Cepheids. Instead of drawing a contour diagram of mean amplitudes, we draw contour diagrams of the frequencies of amplitudes less than $0^m.7$ and greater than $1^m.0$ (Fig. 13). The picture is now greatly simplified, and suggests that the complexities of Fig. 9 are a consequence of two overlapping distributions. This possibility cannot be profitably explored further until accurate photoelectric amplitudes are available.

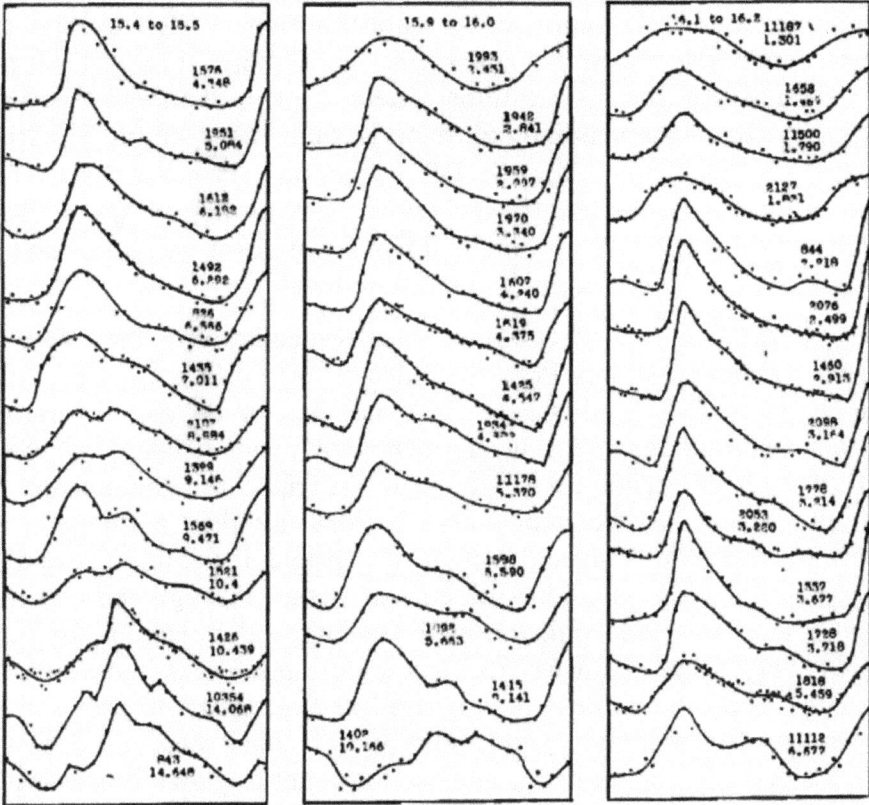

Fig. 11.—Light curves of Cepheids in the Small Magellanic Cloud, mean apparent photographic magnitudes 15.4, 15.9, and 16.1. Arrangement as in Fig. 10.

Finally, let us compare the frequency of periods of stars in different parts of the period-luminosity plane. The results derived from over 600 Cepheids in the Small Cloud are shown as a contour diagram in Fig. 14. The division into two groups is very evident. If, again, we suppose that a Cepheid moves from left to right across the diagram, then we can infer that its progress is not uniform.

The rate of development might be expected to be a function of the mass (*i.e.*, the luminosity), faster for more luminous stars. It should also be a function of the time: for constant mass, log P should increase as 3/2 log R, where R is the star's radius. Thus, if the radius increased in proportion to the time, log P should increase at an accelerated rate, and the greatest number of Cepheids at any one luminosity should occur for the shortest periods. However, it seems probable from analogy with the earlier development of massive stars, that R itself increases at an accelerated rate with time, so the concentration of Cepheids at the shortest periods would be enhanced even more. Fig. 14 is not consistent with these expectations. There are two maxima of period-frequency for all stars at and below magnitude 16. A similar situation has already been noted for the RR Lyrae stars in ω Centauri. The observations indicate that if a Cepheid moves across the period-luminosity plane, its development slows down in the neighborhood of two periods, different for each magnitude level.

If the two sets of contours refer respectively to the overtone and the fundamental, however, they also refer to different values of R at a given period. In this case, in order to convert Fig. 14 into a contour diagram representing frequency of mean density, we must shift the lower-period distribution to the right by an appropriate amount. If the two frequency distributions are thus combined, we obtain Fig. 15, which strongly suggests a single distribution, and also raises the question whether a given star passes through the Cepheid domain with overtone pulsation, or fundamental pulsation, or both successively. Here again we need accurate light curves and colors in order to clarify the picture. A diagram like Fig. 15 might prove to be the most fundamental representation of the relation between luminosity and mean density, and we note that its slope in the lower part is greater than that of the period-luminosity relation.

The question of the rate at which an individual Cepheid traverses the HR plane leads immediately into the wider question of the significance of the frequency of periods. The large number of Cepheids of short period in the Small Cloud, as compared with our own galaxy, has been discussed by Shapley and McKibben (40), who regard the difference as real. Indeed, it is difficult to suppose that selection and obscuration could have cut down the numbers of galactic Cepheids with periods less than three days by a factor of ten; down to a given apparent magnitude, the Cepheids of longer period are at a greater disadvantage.

82

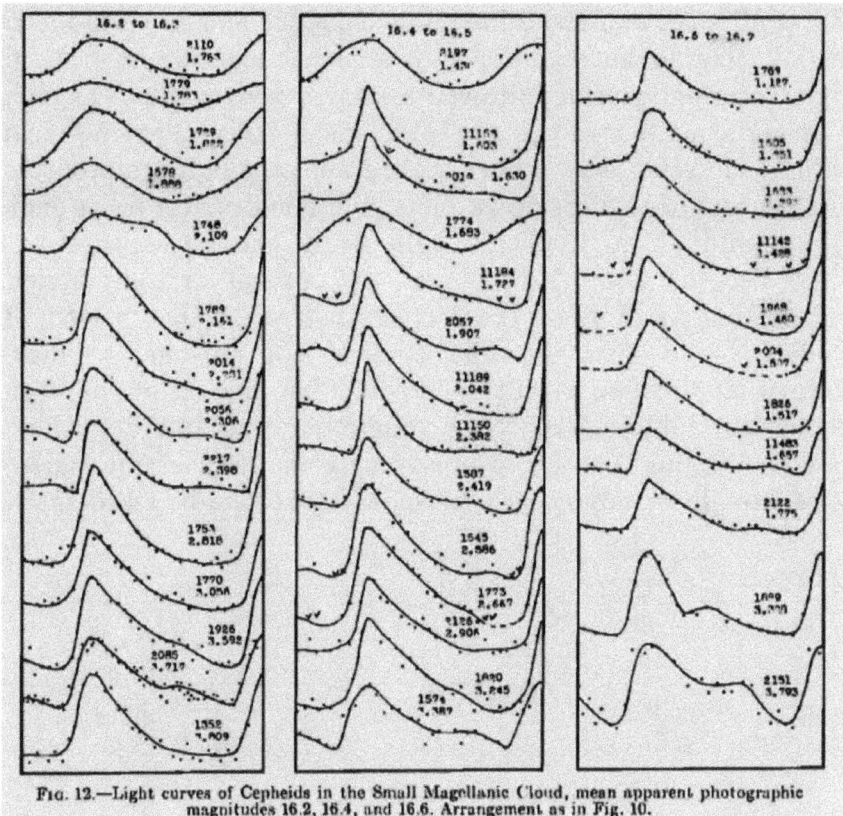

Fig. 12.—Light curves of Cepheids in the Small Magellanic Cloud, mean apparent photographic magnitudes 16.2, 16.4, and 16.6. Arrangement as in Fig. 10.

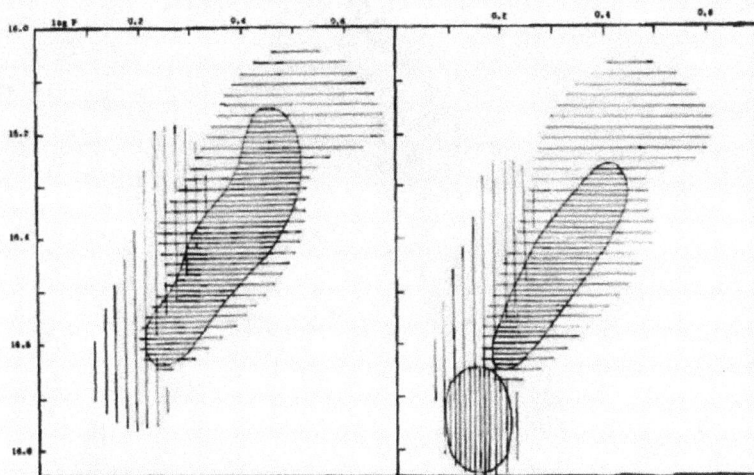

Fig. 13.—Distributions of amplitudes for Cepheids of average photographic magnitude 16.0 and fainter in the Small Magellanic Cloud. Left side: light shading, over 4 stars between limits of 0.1 in log P and in m_{pg}; heavy shading, over 6 stars. Horizontal shading: amplitude over $1^m.0$; vertical shading: amplitude less than $0^m.7$. Right side: light shading: over 6 per cent of Cepheids in given magnitude interval; heavy shading: over 10 per cent of Cepheids in given magnitude interval. Horizontal and vertical shading have the same meaning as on the left.

If all stars that leave the main sequence and move to the right in the HR diagram become Cepheids at some stage, we might (if their progress was uniform or followed a known law) predict the number of Cepheids per cubic parsec by means of the Salpeter "creation function." However, as we have argued above, a uniform progression does not seem to fit the known facts. We could invert the argument, and, assuming that Cepheids follow the period-luminosity law exactly, use the number of observed Cepheids at each period to calculate the duration of the Cepheid stage at that period. This procedure is so uncertain that I shall not presume to make a numerical application of it; it predicts an increasing number of Cepheids per cubic parsec down to periods of about a day, since all stars down to this luminosity presumably rate as massive stars, and may be thought to move nearly horizontally across the HR diagram.

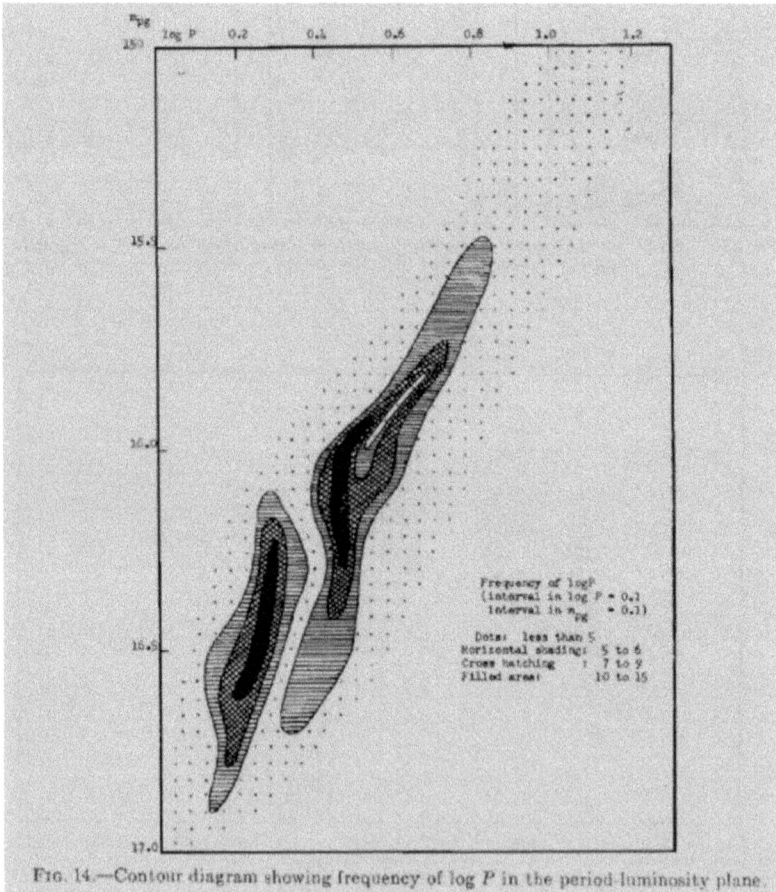

FIG. 14.—Contour diagram showing frequency of log P in the period-luminosity plane.

84

However, the data can be used in another way, to examine the source of the difference between the period distribution in the Small Cloud and in our own galaxy. If the Cepheids are strictly comparable in the two systems, their rate of progress at a given absolute magnitude should be the same, and if the distribution of periods is as different as it appears to be, the only conclusion that can be drawn is that the "birth function" in the Small Cloud differs from that in our galaxy, and increases more steeply with decreasing luminosity, at least down to absolute magnitude −1.5. Information on the observed luminosity function in the Small Cloud is very indefinite, but this is an observational datum that could readily be obtained. It is possible, as Arp has suggested, that Cepheids in the Small Cloud (and, by inference, the Small Cloud itself) may differ from their galactic counterparts in chemical composition. A difference in chemical composition would, if large enough, have a noticeable effect on the mass-luminosity relation (41), and perhaps on the luminosity function itself.

In conclusion, I can summarize my opinion concerning the present status of the period-luminosity relation. The preliminary task of determining its average course has been completed, and we are nearing a consensus concerning the colors of Cepheids, and the relation of mean color to period. Future work must be concerned with the dispersions of the period-luminosity relation, the period-luminosity-color relation, and the period-luminosity-light curve relation. The part played by the Cepheid stage in stellar development will stimulate studies of the relation of period-frequency to the local birth function and to the rate of progress of a star through the variable stage. These last problems are intimately tied up with the study of stellar interiors, and the machine computation of evolutionary tracks. In this area, speculation is worse than valueless, and the greatest service that can be rendered by the student of variable stars is the provision of data that are accurate and complete—in other words, the systematic discovery of variable stars in carefully selected systems, accurate studies of brightness, light curve and color, and determination of luminosity functions.

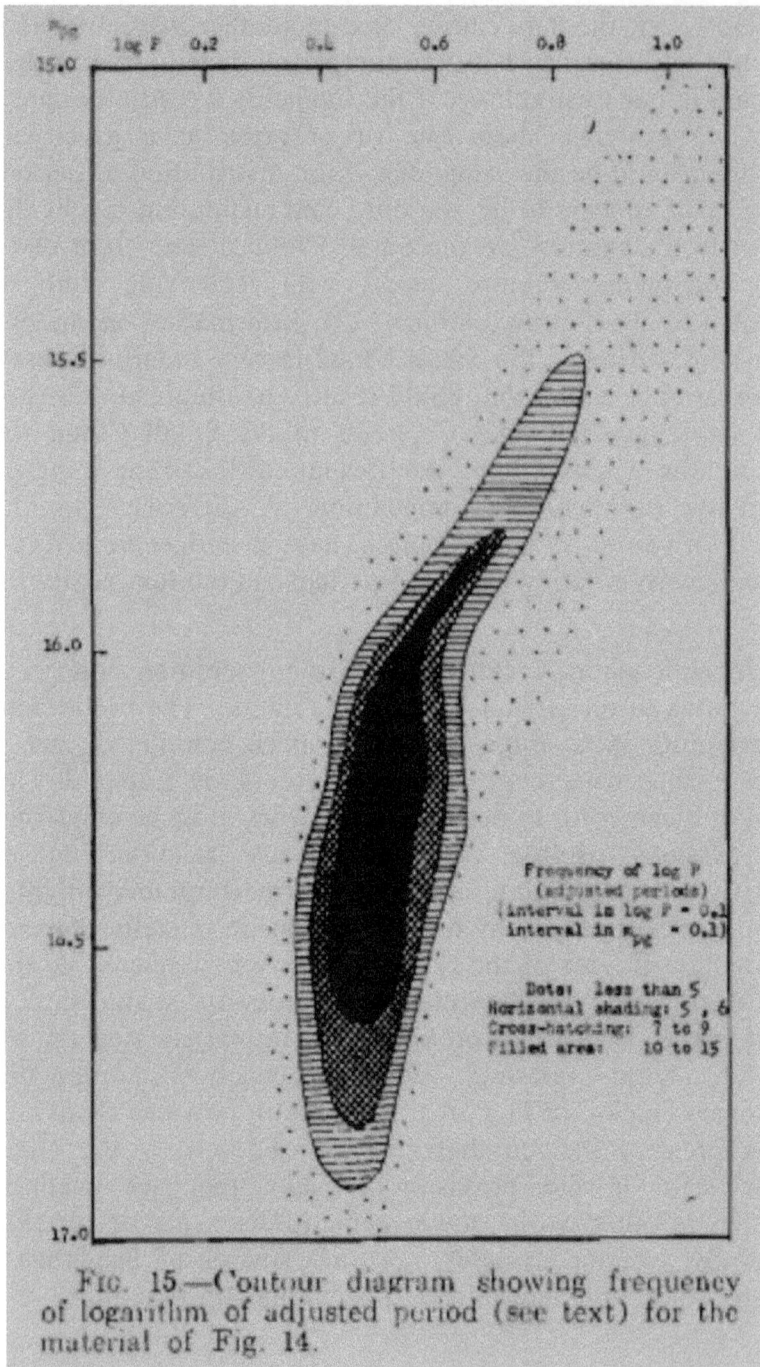

FIG. 15—Contour diagram showing frequency of logarithm of adjusted period (see text) for the material of Fig. 14.

REFERENCES

1. PICKERING, E. C. Circ. Astr. Observ. Harvard Coll. 173. 1912.
2. BAILEY, S. I. Ann. Astr. Observ. Harvard Coll. 38. 1902.
3. SHAPLEY, H. *Star clusters:* 125. 1930.
4. HUBBLE, E. P. Astrophys. Journ. **62:** 409. 1925; **63:** 17. 1926; **69:** 145. 1929.
5. BAADE, W. Trans. Int. Astr. Union 8: 397. 1954; Publ. Astr. Soc. Pacific **68:** 1. 1956.
6. ROBERTS, M., and SANDAGE, A. R. Astr. Journ. **59:** 190. 1954.
7. SANDAGE, A. R. *Stellar populations:* 52. Vatican Observatory, 1958.
8. ARP, H. C. Handb. der Phys. **51:** 118. 1958.
9. HERTZSPRUNG, E. Bull. Astr. Inst. Netherlands 96. 1926.
10. PAYNE-GAPOSCHKIN, C. Ann. Astr. Observ. Harvard Coll. **113:** 173. 1954.
11. SHAPLEY, H., and NAIL, V. McK. Proc. Amer. Phil. Soc, **92:** 310, 1948.
12. PAYNE-GASPOSCHKIN, C. *Variable stars and galactic structure:* 42. 1954.
13. PAYNE-GAPOSCHKIN, C. Astr. Journ. **52:** 218.1947.
14. PAYNE-GAPOSCHKIN, C. *Vistas in astronomy:* 1142. 1956.
15. PAYNE-GAPOSCHKIN, C. *Variable stars and galactic structure:* 13. 1954.
16. SHAPLEY, H. *Star clusters:* 132. 1930.
17. ARP, H. C. Astr. Journ. **63:** 45. 1958.
18. SANDAGE, A. R. Astrophys. Journ. **127:** 513. 1958.
19. CODE, A. Astrophys. Journ. **106:** 309. 1947.
20. WALRAVEN, TH., MULLER, A. B., and °OOSTERHOFF, P. TH. Bull. Astr. Inst. Netherlands 484, 1958.
21. GASCOIGNE, S. C. B., and KRON, G. E. Publ. Astr, Soc. Pacific **64:** 196. 1952.
22. GASCOIGNE, S. C. B. Australian Journ. Sci. Suppl. 1957.
23. GASCOIGNE, S. C. B., and EGGEN, O. J. Monthly Notices Roy. Astr. Soc. **117:** 423. 1957.
24. EGGEN, O. J. Astrophys. Journ. **113:** 367. 1951.
25. EGGEN, O. J., GASCOIGNE, S. C. B., and BURR, E. J. Monthly Notices Roy. Astr. Soc. **117:** 423. 1957.
26. PAYNE-GAPOSCHKIN, C. *Variable stars and galactic structure:* 34. 1952.
27. BURBIDGE, G. R., and BURBIDGE, E. M. Handb. der Phys. **51:** 186.

1958.

28. SANDAGE, A. R. Astrophys. Journ. **126:** 326. 1957.

29. ARP, H. C. Handb. der Phys. **51:** 110. 1958.

30. ABT, H. A. Astrophys. Journ. **126:** 138. 1957.

31. IRWIN, J. B. Proc. Nat. Sci. Found., Charlottesville. 1956.
KRAFT, R. P. Astrophys. Journ. **126:** 225. 1957; **128:** 161. 1958.
ARP, H. C. Handb. der Phys. **51:** 99. 1958; Astrophys. Journ. **128:** 166. 1958.
SANDAGE, A. R. Astrophys. Journ. **128:** 150. 1958.

32. STARIKOVA, G. A. Variable Stars (U.S.S.R.) **7:** 124. 1949.

33. BURBIDGE, G. R., and BURBIDGE, E. M. Handb. der Phys. **51:** 208. 1958.

34. IRWIN, J. B. Proc. Nat. Sci. Found., Charlottesville. 1956.

35. SANDAGE. A. R. *Stellar populations:* 41. Vatican Observatory, 1958.

36. HOERNER, S. VON. Zeitschr. für Astrophys. **42:** 273. 1957.

37. JOHNSON, H. L. Astrophys. Journ. **126:** 134. 1957.

38. THACKERAY, A. D. *Stellar populations:* 195. Vatican. Observatory, 1958.

39. SHAPLEY, H. and NAIL, V. McK. Astr. Journ. **55:** 249. 1951.

40. SHAPLEY, H., and NAIL, V. McK. Proc. Nat. Acad. Sci. **26:** 105. 1940.

41. HOYLE, F. *Stellar populations:* 223. Vatican Observatory, 1958.

Introduction to

The Advanced X-Ray Astrophysics Facility

Martin Weisskopf is the NASA Project Scientist for the Chandra X-Ray Observatory. The following article describes the justification and initial design for a space based x-ray telescope called AXAF (for Advanced X-Ray Facility). It appeared in 1981 in the Washington Academy of Sciences Journal Volume 71 and is one of several articles dedicated to x-ray astronomy appearing in that issue of the Journal.

AXAF was launched July 23, 1999. As is the case with other NASA satellites, after launch the name is changed to honor a worker in the field. It is now called Chandra after Chandrasekhar (1983 Nobelist). One can find current information about the satellite at http://chandra.harvard.edu/.

The Advanced X-Ray Astrophysics Facility

Martin C. Weisskopf

Space Sciences Laboratory NASA/Marshall Space Flight Center
Huntsville, Alabama USA 35812

Introduction

The HEAO-2/Einstein Observatory employed for the first time in satellite X-ray astronomy, the technology of imaging optics in the direct study of galactic and extragalactic X-ray sources. The increase in sensitivity (a factor of 1000 for the detection of point sources) over UHURU only hints at the tremendous advance in observational capability that focusing optics has provided to X-ray astronomy. The results presented by Dr. Tananbaum elsewhere in these proceedings have clearly demonstrated the importance and significance of imaging X-ray optics and the field of X-ray astronomy. It is because of the success of the HEA0-2/Einstein, both technical and scientific, that the next major program, the Advanced X-Ray Astrophysics Facility (AXAF), is readily identified and well defined. In what follows I will briefly summarize the planned capability of this observatory and indicate some of the astrophysical problems that AXAF will be well suited to address.

1. The Observatory

The AXAF will be an X-ray observatory built around a large-area, high-resolution, grazing incidence X-ray telescope. Designed to operate in space for 10 to 15 years, the AXAF will be operated as a major national facility with the majority of the observing time set aside for guest investigators. An artist's conception of the AXAF in orbit is shown in Figure 1, and the major elements of the AXAF are identified in Figure 2. Also shown in Figure 1 is the Space Shuttle which will be used to place the AXAF in orbit and revisit the observatory at approximately 3-year intervals for the purpose of refurbishing and/or replacing instruments.

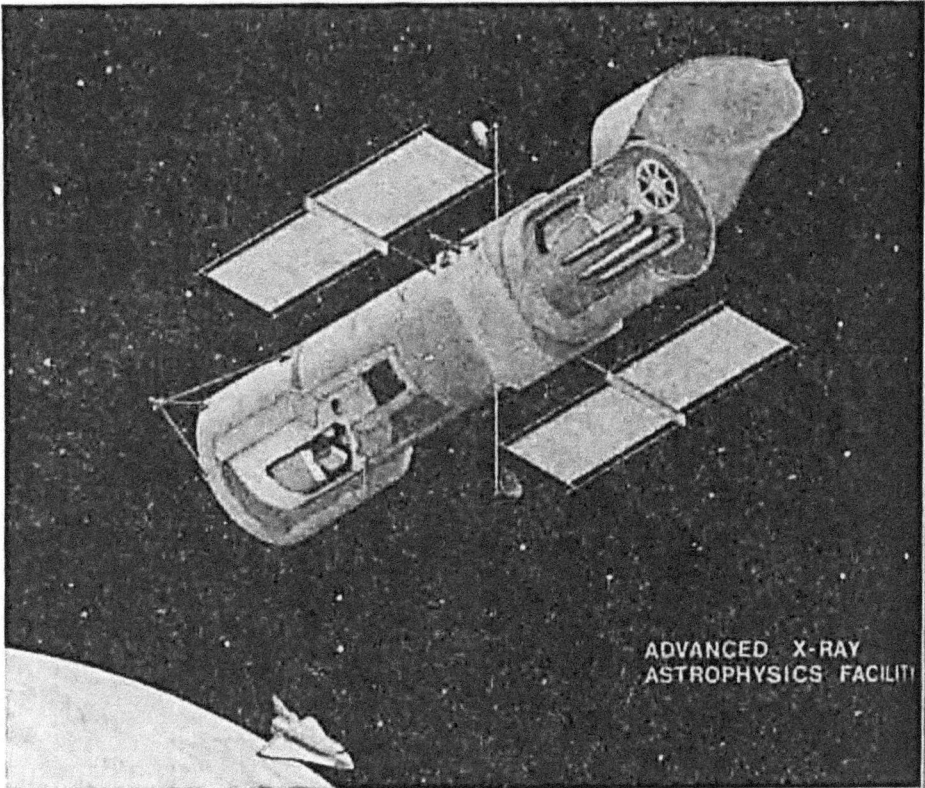

Fig. 1. An artist's conception of the AXAF in orbit.

The long lifetime of AXAF will provide us with a facility not only capable of performing the observations now known to be necessary because of previous investigations and the questions raised by them, but also to follow up, in coordinated observing programs, those new discoveries which AXAF will surely make.

The heart of the AXAF is an X-ray telescope made up of six nested Wolter type I paraboloid-hyperboloid pairs ranging in diameter from 0.6 to 1.2 meters. The geometric collecting area will be 1700 cm^2, and the focal length will be 10 meters. The energy response will extend to well above 8 keV. The telescope will have angular resolution of better than 0.5 arc-second on axis and independent of energy, and a significant (but energy dependent) fraction of the reflected flux within the central core.

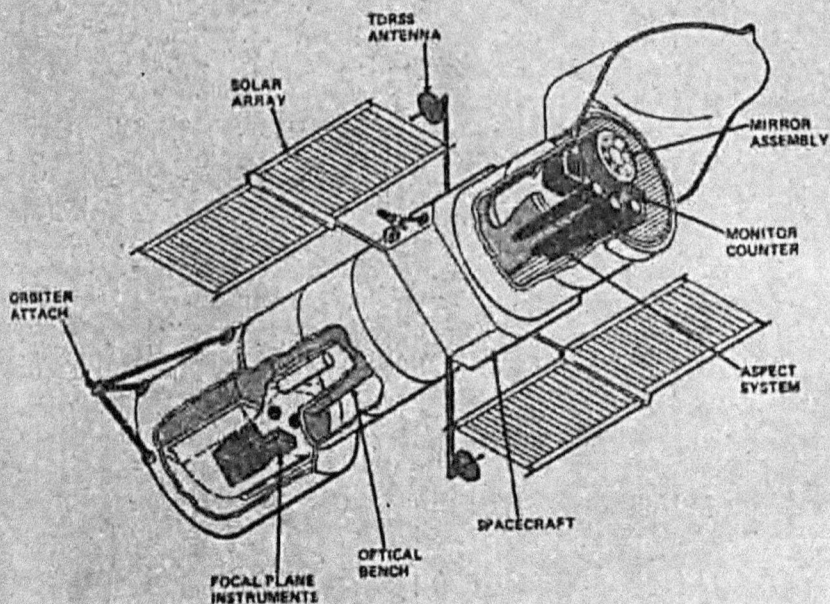

ADVANCED X-RAY ASTROPHYSICS FACILITY

Fig. 2. Cutaway showing the major elements of the AXAF.

The baseline design of the AXAF has evolved from an interaction between the scientific requirements and engineering constraints. It has been enhanced as a result of the experience gained in the design, fabrication, assembly, test and performance of the HEAO-2/Einstein telescope. The AXAF telescope parameters are summarized in Table 1, where they are also compared to those of the Einstein Observatory. There are several important differences between the two telescopes, and these are only partially apparent from the factor four increase in geometric area and the factor eight goal for the improvement in angular resolution listed in the table.

Certainly the larger geometric collecting area will make the AXAF far more efficient than the Einstein telescope over the energy bandwidth that the two observatories have in common. Furthermore, the range of grazing angles permitted by the AXAF design allows the response of the telescope to extend to energies well beyond 7 keV. The total (on-axis) effective collecting areas are compared in Figure 3. The extension of X-ray imaging to the energies shown in the figure

92

will have important consequences. This response includes the complex of lines due to highly ionized iron and permits such studies as the iron line spectroscopy of nearby supernova remnants and also the direct measurement of redshifts from clusters of galaxies where we know, from UHURU and its successors, that emission from highly ionized iron exists. Of equal significance, however, are not the observations in this new wavelength range we can now point to, but the fact that we will have available to us a new region of the spectrum for imaging studies. Both UHURU and especially the HEAO-2/Einstein have demonstrated the importance of increasing the wavelength sensitivity, and one can only speculate now as to what surprises await us.

Table 1 – A Comparison of Telescope Parameters

	HEAO-2/Einstein	AXAF
No. of Elements	4 Nested Pairs	6 Nested Pairs
Outer Diameter (m)	0.58	1.2
Focal Length (m)	3.44	10.0
Geometric Area (cm^2)	460	1700
Inner Grazing Angle (°)	0.68	0.45
Outer Grazing Angle (°)	1.17	0.85
Resolution (arc-sec)	4.0	0.5
Field of View (°)	1	1

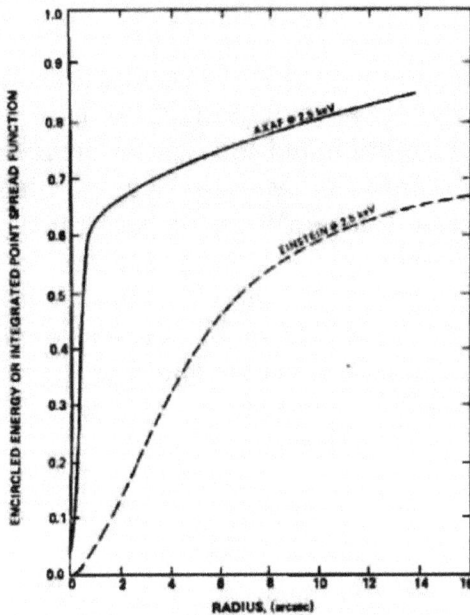

Fig. 3. The AXAF and Einstein on-axis effective areas as a function of energy.

The most important and significant difference between the AXAF and its prototype is the imaging quality. This is illustrated in Figure 4 where the fraction of on-axis reflected flux at 2.5 keV is shown as a function of the radius of a perfect detector. In general, the imaging performance of an X-ray telescope is not limited by diffraction but by the geometrical figure, the alignment, and the surface finish of the reflecting elements. The figure and alignment determine the full-width-half-maximum of the response function and is independent of energy. The surface finish, primarily microscopic surface roughness, determines the fraction of the reflected flux that remains within any given radius about the center of an X-ray image. Thus, figure and alignment establish the rapidly rising portions of the response curves shown in Figure 4, whereas the amount of scatter from the reflecting surfaces limits the efficiency and the ability to resolve low contrast features. Figure 5 demonstrates the anticipated AXAF performance as a function of energy for resolution elements of 1 and 20 arc-second diameters. The significant AXAF imaging quality over that achieved by the HEAO-2/Einstein is based, for the most part, on a modest extension of the Einstein tolerances and a much better understanding of the contributions to X-ray scatter that have taken place in the years since the HEAO-2/Einstein was fabricated.

Before I discuss the types of instruments one might expect to see aboard the AXAF, it is important to note that the AXAF's improved angular resolution and low scatter are absolutely necessary to attack a large number of astrophysical problems which simply cannot be done at the Einstein level of performance. These include, for example, the "weighing" of X-ray sources in globular clusters where even a three solar mass binary system could be expected to be no more than 1 arc-second from the center of a typical centrally condensed cluster. Both low scatter and high angular resolution are musts for the search for, and study of, low contrast features such as jets near nuclei of active galaxies.

94

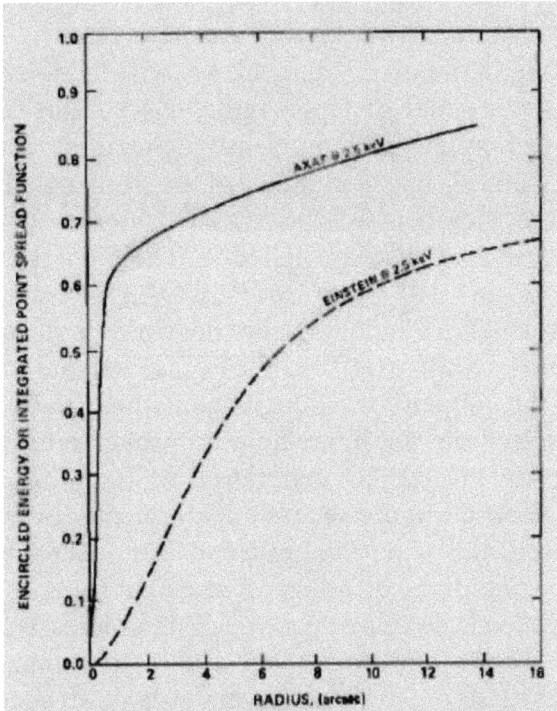

Fig. 4. The fraction of flux within a resolution element of radius R as a function of R at 2.5 keV.

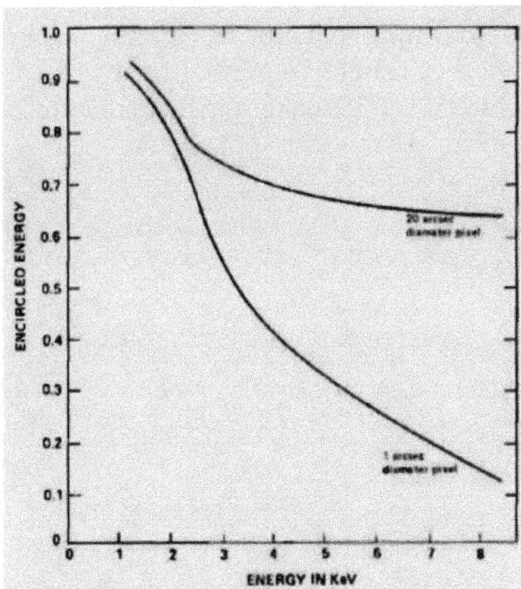

Fig. 5. The fraction of reflected flux within resolution elements of 1 and 20 arc-seconds diameter as a function of energy.

2. Instrumentation

The instrumentation for the AXAF will be selected through a series of Announcements of Opportunity (AO), and the AO for the first instrument complement is currently scheduled to be released in 1981. Tables 2 and 3 list a number of instruments that have been identified by the members of the AXAF Science Working Group. This group, whose members are listed in Table 4 [at end of article], was established in 1977 to advise NASA on the scientific requirements for AXAF and on the possible instrumentation that might be flown. I won't dwell on these instruments except to point out that, first, the longer focal length and, thus, the increased plate scale have resulted in an immediate improvement in the angular resolution of even the HEAO generation of imaging detectors by a factor of three. Second, more recent developments in instrumentation technology offer the exciting prospect of performing spatially resolved, high quantum efficiency, reasonable energy resolution spectroscopy. Indeed, detailed spectroscopic studies with AXAF will encompass a wide range of objectives, and instruments, such as those listed in Tables 2 and 3, will be well suited to accomplish these studies. For example, with an objective grating and a high resolution imaging detector in the focal plane, the spectra of faint compact sources could be obtained with an energy resolution ($E/\Delta E$) of order 100 and, thanks to improvements in grating technology and the AXAF's collecting area, effective areas a factor of a hundred or more over the equivalent HEAO-2/Einstein instrument are possible.

Table 2 – Imaging x-ray Detectors

Detector	Size (Field of View)	Spatial Resolution	Quantum Efficiency	E/ΔE @1 keV	E/ΔE @ 6 keV
Charge Coupled Device	>25mm × 25 mm (≥8 × 8 arc min)	15-25μm (0.3-0.5 arc sec)	high	5	30
Negative Electron Affinity detector	>25 mm diameter ≥8 arc min	15μm (0.3 arc sec)	high	5	1
Microchannel Plate	≥85 mm diameter (≥30 arc min)	15μm (0.3 arc sec)	Low	none	none
Imaging Proportional Counter	180 mm × 180 mm (1°×1°)	< 0.5 mm (<10 arc sec)	high	2	5
Gas scintillating imaging proportional counter	180 mm × 180 mm (1°×1°)	< 0.5 mm (<10 arc sec)	high	10	12

Table 3 X-ray spectrometers and polarimeters

Instrument	Energy Range (keV)	Energy resolution (E/ΔE)	Location
Filter Wheel	0.1–8	3	In front of detector
Transmission Grating	0.1–4	100–200	Behind mirrors
Solid State Detector	0.4–8	15–50	In focal plane
Objective Crystal	0.5–8	500–10,000	In front of mirrors
Focal Plane Crystal	0.5–8	500–3,000	In focal plane
Focal Plane Concave Grating Spectrometer	0.1–1	500	In focal plane
Gas Scintillation Proportional Counter	0.1–8	10–15	In focal plane
Bragg Crystal Polarimeter	2.6, 5.2	N/A	In focal plane

Figure 6 shows the potential sensitivity of AXAF for HEAO2/Einstein of point sources as compared to the HEAO-2/Einstein performance with the High Resolution Imager (HRI) and the Imaging Proportional Counter (IPC) in the focal plane. The AXAF sensitivity is based on the use of a high-resolution imaging detector with the efficiency of a charge-coupled solid state detector. The sensitivity shown in Figure 6 is placed in perspective in Table 5, which lists the minimum detectable luminosity as a function of the luminosity-distance for several characteristic distances and objects. The usefulness of this sensitivity is indicated in Table 6, which lists the known luminosity of several categories of X-ray sources.

Such sensitivity clearly opens the door for many avenues of research. Consider, for example, the study of normal stars. Based on the (unexpected) Einstein results, we now know that stars throughout the H-R diagram will lie well within AXAF's observational capability. By the end of its mission the HEAO2/Einstein will have observed only approximately 500 stars and, of these, probably less than half of the observations will have obtained even low resolution spectra with the IPC. Thus, only statistically limited spectral classifications will be completed, and one will be restricted, for the most part, to relatively crude correlation studies with optical data. Taking the one-hundred-fold increase in the sensitivity of AXAF into account, all of the more than 6000 stars in principle accessible to the Einstein (but not observed) could be studied in a relatively short time. The total number of stars accessible to AXAF is, of course, vastly larger, and the available volume of space for sampling will increase by a factor of a thousand.

The promise of AXAF is perhaps more colorfully illustrated in Figure 7 which shows the HEAO-2/Einstein images of M3I. The AXAF would add a third image, further resolving the galactic nucleus and, in clarity, would be to the HRI image is to the IPC image. Furthermore, with the AXAF the study of discrete sources in normal galaxies will be extended to objects with much lower luminosity than was possible with HEAO-2/Einstein. Sources as weak as 5×10^{34} ergs/s could be detected in Andromeda, and the coronae of O and B stars would be visible in the Magellanic Clouds. The study of individual high-luminosity (10^{37} ergs/s) sources, now only possible for M31, will be extended to galaxies at distances up to 20 Mpc. Thus, all 2500 galaxies in the Virgo Cluster will be amenable to the type of observation illustrated in Figure 7. With such data one will be able to study the spatial and luminosity distributions with an angular resolution which can distinguish between bulge, disc and nuclear components and can correlate these distributions to galaxy properties: morphology, metallicity, *etc*. Insofar as the integrated emission of galaxies is concerned, the sensitivity of AXAF will increase the sample from approximately 1 normal galaxy/1° field to as many as 1000/1° field. Among other things, these observations can be used to determine whether high-luminosity activity in the galactic nucleus is confined to a small fraction of all galaxies or whether all galaxies spend a small fraction of their life in a highly active state. Clearly such observations can also shed some light on the relationship between normal and active galaxies.

Another area of research in which the AXAF is prepared to provide unique and significant advances is in the study of clusters of galaxies. This follows directly from the existence of the hot intra-cluster gas first detected by UHURU; the importance of this gas has been emphasized by many of the results presented earlier today. X-ray imaging with the HEAO2/Einstein has begun the detailed study of cluster morphology, so important since the gas is invisible at longer wavelengths. Yet, these observations are sensitive only to the cooler gas because of a limited energy response. Furthermore, detailed mapping and spectroscopy with the HEAO2/Einstein are limited by the spatial sensitivity. With the AXAF, spectrally resolved high-resolution maps of a statistically significant sample of clusters will be possible. Based on a 6-month observing program and an assumed density of 10^{-6} cluster sources/Mpc^3 (L_x 10^{44} ergs/s), 800 sources could be detected in a survey of one hundred square degrees. Moreover, the AXAF should allow for: (1) the detection of the integrated emission from clusters at redshifts up to one and possibly as large as four, depending on their epoch of formation and early evolution; (2) the detection and spatial resolution for clusters as distant as $Z = 1$ to 3; and (3) the detailed mapping and spectroscopy, including x-ray measurements of the redshift for Z as large as 0.5 to 1. Observations such as these will allow for the detailed study of the formation and evolution of the clusters, the origin and heating mechanisms of the intra-cluster medium and the matter content, with emphasis on the hot gas. X-ray observations of clusters with the AXAF should also provide the basis for at least two tests of cosmological models: the first through the differential number counts of X-ray clusters at large redshifts and the second (described in more detail by Dr. Fabian elsewhere in these proceedings) combining X-ray and microwave measurements to determine the deacceleration parameter in an almost assumption-independent way.

Conclusions

I have only touched upon the capabilities of the AXAF and some of the types of studies which could be carried out with this observatory. A far more detailed description of the AXAF, its scientific objectives and its potential for accomplishing these objectives may be found in the report of the AXAF Science Working Group (NASA TM-78285, May 1980).

In conclusion, I believe it is adamantly clear from the results we have heard today that X-ray astronomy has come of age. The tremendous success of the last X-ray astronomy mission, the HEAO2/Einstein, is underscored by the far greater number of astrophysical questions which now confront us, as opposed to what we knew before that mission began. An essential point in charting the future course of our field is that AXAF is not only necessary to perform the required observational tasks to answer these questions, but that AXAF alone has the capability to carry out the majority of the major research goals before us. With the AXAF, X-ray astronomy will be able to take its place, along with radio and optical astronomy, as one of the major probes of the Universe.

Table 4—Members of the AXAF Science Working Group

Professor Riccardo Giacconi—Harvard University—Chairman
Dr. Martin C. Weisskopf—Marshall Space Flight Center—Vice Chairman
Dr. Elihu Boldt—Goddard Space Flight Center
Professor Stuart Bowyer—University of California, Berkeley
Professor George Clark—Massachusetts Institute of Technology
Professor Arthur Davidson—Johns Hopkins University
Professor Gordon Garmire—Califomia Institute of Technology
Professor William Kraushaar—University of Wisconsin
Professor Robert Novick—Columbia University
Dr. Albert Opp—NASA Headquarters—ex officio
Professor Minoru Oda—Tokyo University, Japan
Professor Kenneth Pounds—University of Leicester, United Kingdom
Dr. Seth Shulman—Naval Research Laboratory
Dr. Harvey Tananbaum—Harvard/Smithsonian Center for Astrophysics
Dr. Joachim Truemper, Max-Planck Institute, Germany
Professor Arthur Walker—Stanford University

Table 5—Minimum Detectable Luminosities for 10^5 Seconds of Observation.			
Luminosity-Distance	Object	L_{min} (ergs/s)	Linear-Dimension (per arc-second)
150 pc	star	10^{27}	50 Au
0.7 Mpc	point source in M3I	2×10^{34}	3.5 pc
19 Mpc	point source in Virgo Cluster	10^{37}	100 pc
300 Mpc	normal spiral galaxy	3×10^{39}	1.5 Kpc
1000 Mpc	active galaxy in Hydra	3×10^{40}	5 Kpc
2×10^3 Mpc	quasar	10^{45}	1 Mpc

Table 6 Examples of Known X-ray Luminosities

Object	Luminosity (ergs/s)
Sun	5×10^{27}
Cen X-3	2×10^{37}
Milky Way	5×10^{39}
M87	3×10^{43}
3C273	5×10^{45}

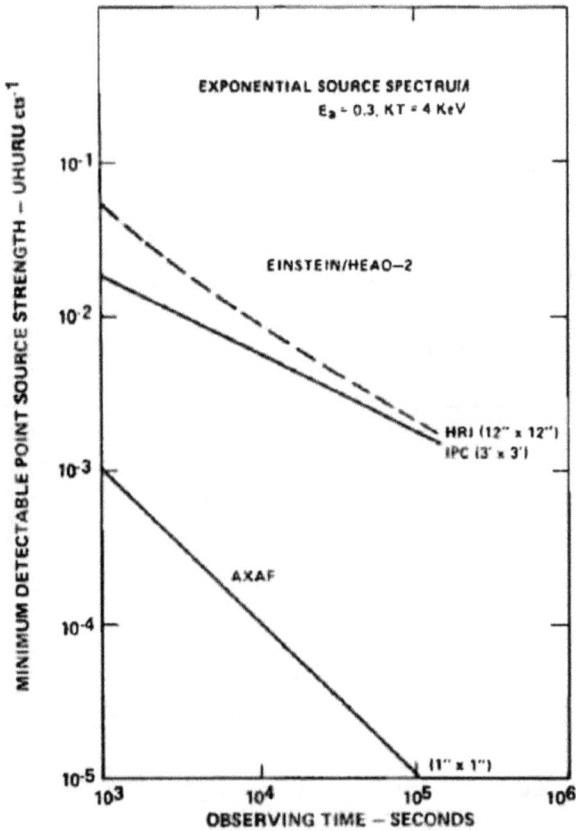

Fig. 6. The sensitivity of the AXAF for the detection of point sources as a function of observing time as compared to the Einstein Observatory. The calculation is based on a moderately efficient charge-coupled imaging detector in the focal plane.

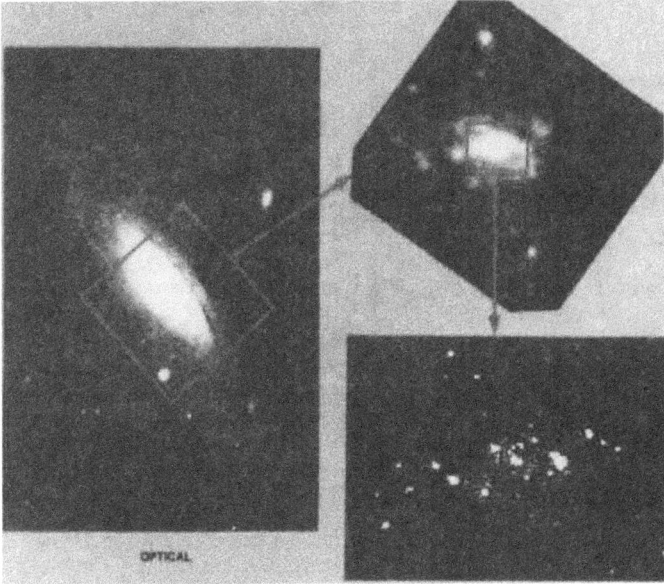

Fig. 7. Optical and X-ray picture of M31 (Andromeda). IPC observation (upper right). HRI observation (lower right)

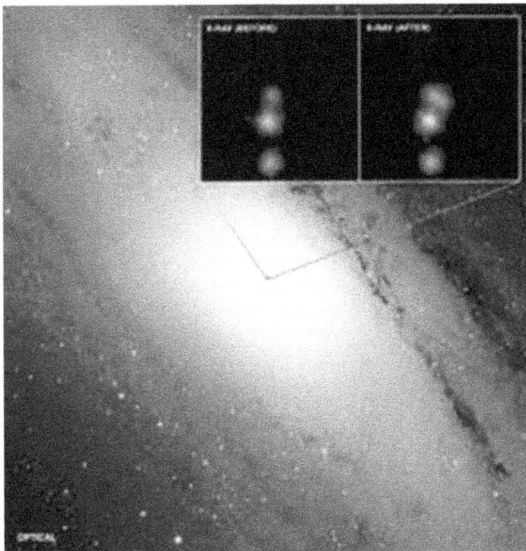

[Ed] Figure – inserts taken by Chandra superimposed on an optical image of M31

Introduction to

The Galaxy No One Wanted to See

Gene Byrd is a professor emeritus at the University of Alabama; Sethanne Howard is the retired chief of the Nautical Almanac Office. The following article appeared in 2007 in the Journal of the Washington Academy of Sciences, Volume 93. The authors discuss the strange galaxy NGC 4622. Long thought to be a normal galaxy, the authors show that it is one of the few galaxies that have opposite rotating arms; *i.e.*, spiral arms going in both directions: counterclockwise and clockwise. In general, it is believed that spiral arms in a galaxy spiral only in one direction. The letters NGC stand for New General Catalog, and the galaxy is number 4622 in that catalog.

The Galaxy No One Wanted To See

Gene G. Byrd and Sethanne Howard
University of Alabama, USNO (retired)

Abstract

NGC 4622 is an intriguing galaxy. Byrd, Buta, and Freeman (2003) found that its strong lop-sided pair of outer arms is leading in the clockwise rotating disk. It has a weak single inner trailing arm that nonetheless lasts through 520°. This runs counter to accepted theory which assumes that all spirals have outer trailing arms. This galaxy is a problem calling out for an explanation. The VBI (Visual/Blue/Infrared) data provide an independent determination that the inner single arm trails and that the outer pair leads. Fourier decomposition confirms the result.

Introduction

Spiral galaxies are quite common in the Universe. They appear to be disks in space, often described as having a fried egg shape. When seen from the side, they are remarkably thin compared to their extent by a factor of at least one to fifty; *i.e.*, the diameter of the disk is more than fifty times the thickness of the disk. They also tend to be flat, occasionally (rarely) tilting up and down at the edges like the brim of a fedora hat. Most have a bulge at the center (the yolk of the egg). Although they are disk like in appearance they are not plain; they have a wide variety of internal structure and form some of the most striking objects in the sky. Described as spiral arms, these structures can be as simple as a pair of beautiful spiral arms (*e.g.*, M51, Figure 1) or as complex as a multi-armed galaxy (*e.g.*, M99, Figure 1). They can have two, three, four, or multi-arms; they can be flocculent in nature. Some spiral galaxies have a bar shape that stretches across the middle.

Although, these galaxies, when viewed from the side, appear very thin; most of them probably have a more spherical 'halo' of "dark matter" surrounding the disk. This halo is said to be dark matter because it emits no light but exerts a gravitational force like ordinary matter. Computer simulations support this theory because a perturbed rotating stellar disk without a stabilizing halo of some mass will fragment or become chaotic.

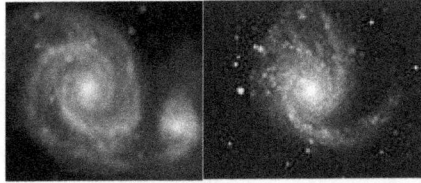

Figure 1 - M51 (left) and M99 (right)

The winding of spiral arms is an important dynamical and kinematic feature of any spiral galaxy. Whether they trail or lead the direction of rotation of the disk has been of interest since spirals were first noticed. Figure 2 is a schematic of CW rotating disks with trailing and with leading arms. Trailing is defined as winding outward opposite the sense of disk orbital motion; leading is winding outward in the same sense as the disk orbital motion.

Figure 2

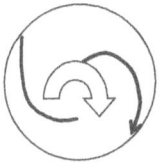

leading

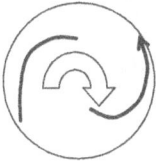

trailing

Historically, astronomers took both viewpoints. Based on theoretical studies B. Lindblad (1941) said spiral arms led. E. Hubble (1943) said they trailed. Based on kinematic studies where one can unambiguously distinguish the near side of the disk, G. de Vaucouleurs (1958) determined that they trailed. He used an asymmetry in the dust distribution to determine which side of the minor axis is the near side. The dust asymmetry is caused by the fact that in an inclined galaxy, dust in the near side is silhouetted against the background starlight of the bulge and disk. This determination combined with the arm winding sense outward CW or CCW on the sky plus Doppler shift observations permitted verification that the arms trail. In galaxies with significant nearly spherical bulge components, this effect can be seen even if the galaxy is more nearly face-on. In their discussion of density wave theory, Binney and Tremaine (1987) argue that leading arms are not likely to be seen because they would quickly unwind and become trailing arms.

Theoretical studies indicate that some tidal encounters (where a companion galaxy passes nearby the primary) can generate a leading arm. Computer simulations support the theory by showing that retrograde encounters in the presence of a large halo-to-disk mass ratio can produce a leading arm. Yet, although leading arms do have a theoretical basis, the general consensus still has been that spiral arms trail.

Initial Work On NGC 4622

NGC 4622 is a spiral galaxy which at first glance appears to have a deceptively normal appearance. Figure 3 shows a Hubble Space Telescope photo of it. However, in a blue sensitive, ground based image of NGC 4622, Byrd, *et al* (1989) pointed out that in addition to the pair of strong, lopsided outer arms winding outward CW, NGC 4622 has a weaker, single inner arm winding outward CCW inside a ring. In other words, NGC4622 has two nested *oppositely* winding spiral arms. Thus either the inner single arm or the outer pair of arms of NGC4622 must be leading.

Figure 3 – NGC 4622

NGC 4622 is a southern, ringed spiral galaxy in the Centaurus cluster putting it about 40 Mpc (Mega-parsecs) or 130 million light years away. As said before, it has two strong CW outer arms wrapping around a large bulge along with the inner single CCW arm. As is typical of such galaxies, the bulge contributes about half of the light. The galaxy is almost face-on with an inclination (tilt from face-on) of about 19° (see Buta, Byrd, and Freeman 2003). Even though it does have this unusual arm pattern, NGC 4622 is not unique. There are at least three other galaxies that show features in common with NGC 4622's spiral pattern; *e.g.,* the Blackeye Galaxy (NGC 4826), ESO 297-27, and NGC 3124.

NGC 4622 provides one of the most convincing cases of a rare leading spiral arm in *any* galaxy. Using multi-band ground-based surface photometry, Buta, Crocker, and Byrd (1992) showed that the single inner arm is a stellar dynamical feature, not the result of a chance dust distribution of young stars or gas. Their ground based photos verified the existence of the inner arm in the stellar disk. So, either the outer arms lead and the inner arm trails, or the outer arms trail and the inner arm leads.

Accepting the general consensus that pairs of arms trail Byrd, Freeman, and Howard (1993) provided an *n*-body simulation which produced a single *leading* inner arm and two outer *trailing* arms. To reproduce the NGC 4622 structure they included a massive halo (eight times the mass of the disk) and a plunging retrograde encounter of a small (1/100 the mass of the disk) perturber. This produced both a single inner and outer pair of arms. In their simulations, they found that a retrograde encounter with a massive perturber cannot form *both* trailing and leading arms in one galaxy. It only takes a small companion to produce the global structure.

Verifying Work On NGC 4622

As a result of the initial discovery in a blue-light ground-based photograph, the ground-based image of the stellar disk, and the simulations, the situation with NGC 4622 was thought to be settled. To double check the conclusions, Buta, Byrd, and Freeman (2003) obtained a set of four new observations: Hubble Space Telescope images in four colors; a Cerro Tololo Inter-American Observatory (CTIO) Fabry-Perot Hα velocity field; CTIO three color images; and Parkes 64m 21-cm radio data. Analysis of the velocity field showed that the kinematic line of nodes of the disk is in position angle +22°.

These new verifying observations provide convincing arguments that the two clockwise outer arms *lead* the CW rotating disk instead of trail. This means, therefore, that the inner arm *trails*. To show this, Buta, Byrd, and Freeman (2003) used an extension of the method that de Vaucouleurs used. De Vaucouleurs used Doppler shifts to determine which half of the major axis of a galaxy is receding from us relative to its center, and then used an asymmetry in the observed dust distribution to determine which side of the minor axis is the near side. The dust asymmetry he used is not intrinsic but is caused by the fact that, in an inclined galaxy, dust in the near side is silhouetted against the background starlight of the bulge and disk. In galaxies with significant nearly spherical bulge components this effect can be seen even if the inclination is less than 45°. See Figure 4 for a schematic of the approach.

What this means is the following. Figure 5 shows a high contrast image of NGC 4622 with the kinematic line of nodes shown as the white line. The image is processed such that the redder the object is the whiter its image is. The bluer the object is the darker the image is. East is to the upper left and north is to the upper right. The field is 1.47' square. One can see that eastward (left) of the axis there are thin white strips indicating dust clouds. Westward of the axis any clouds are much less visible.

This means that the east side of the galaxy is nearer to the observer than the west side. Once we know which side is nearer then we use velocity measurements to show which way the galaxy is rotating. Figure 6 shows a grayscale version of the velocity map of this galaxy. It is not very clear in the grayscale version (galaxy velocity maps are most useful when presented in color); however, there is a well defined line of nodes, marked in black with the upper (north) side moving away from us relative to the disk center and the lower (south) coming toward us. On the left and right of the line of nodes there is no toward or away motion relative to the center. Figure 7 is a diagram combining the arm winding on the sky, the line of nodes, the motions toward and away, and finally the way the disk must turn on the sky. The galaxy turns CW on the sky.

Once we know that the disk is turning CW on the sky, and we know that the outer pair of arms is unwinding CW, then we know that the outer pair of arms *leads*! This means that the inner single arm *trails*.

This result is unpopular because it runs counter to accepted dogma that all spirals have outer trailing arms. The most common comment is "I can't see the inner arm". One straightforward way to address this particular issue uses Fourier analysis.

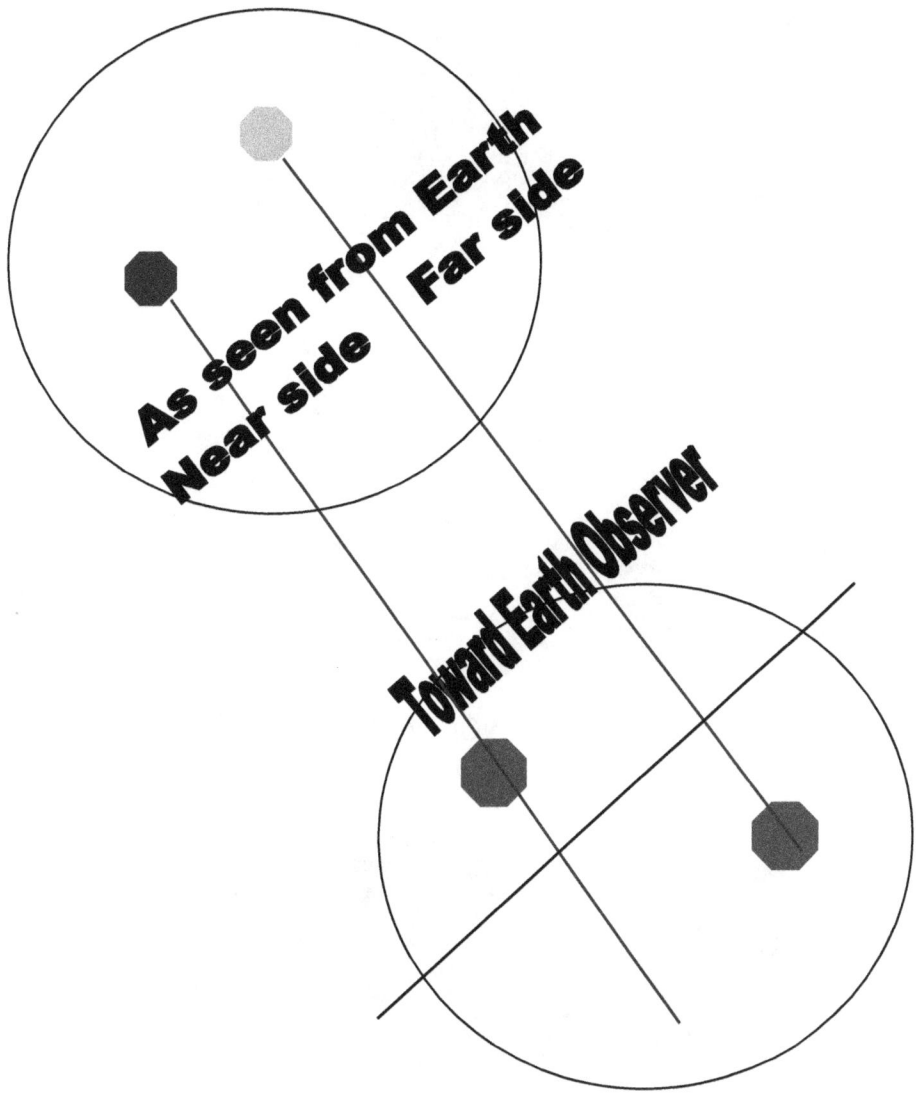

Figure 4 – schematic of the de Vaucouleurs method

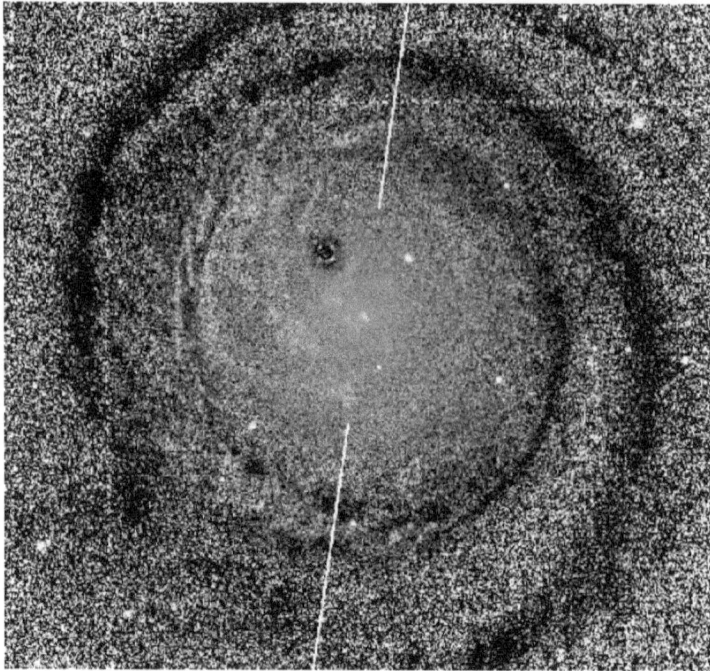

Figure 5 –high contrast image of NGC 4622. Line of nodes is marked in white. The part to the left of the line turns out to be the nearer side.

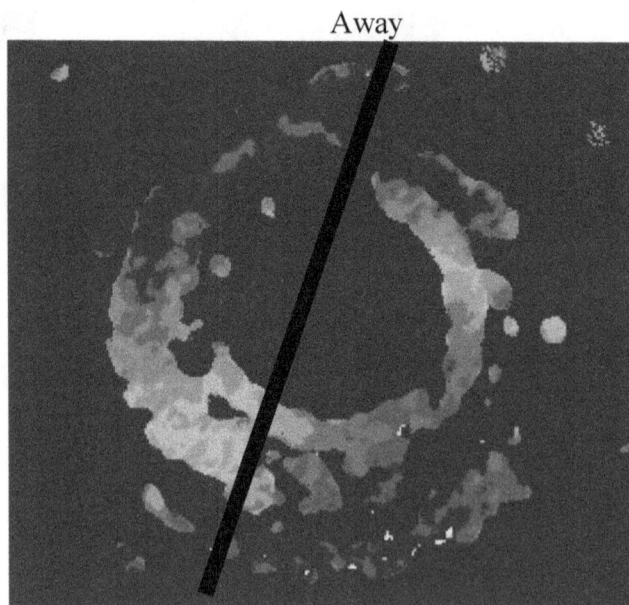

Figure 6 – velocity map of NGC 4622

Which arms lead or trail?

Two arms wind out
in direction of
orbital motion
i.e they lead.

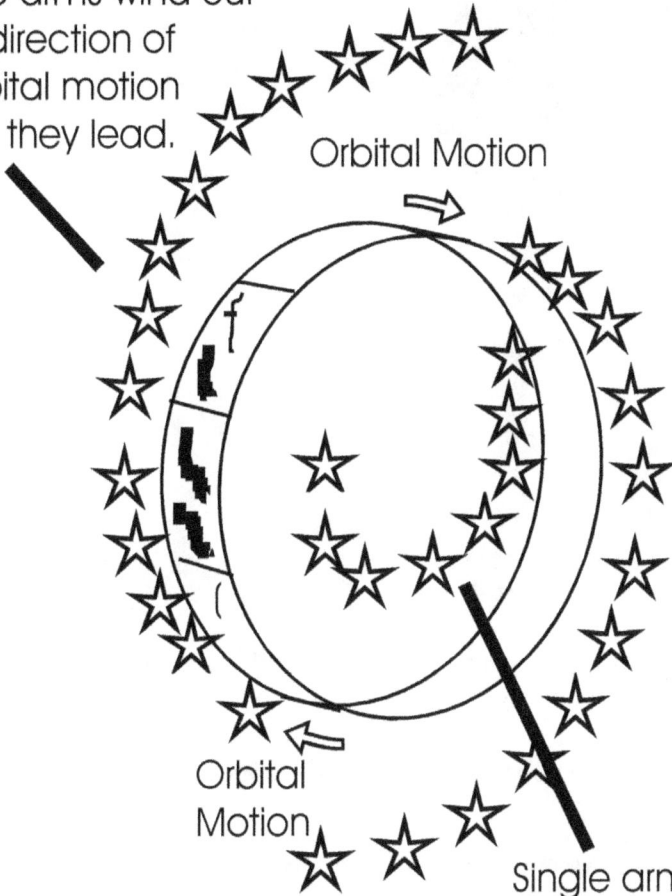

Orbital Motion

Orbital
Motion

Single arm
winds out
opposite orbit
motion i.e. it trails.

Figure 7—diagram of data from photo, dust silhouettes, and Doppler shifts and resulting arm senses.

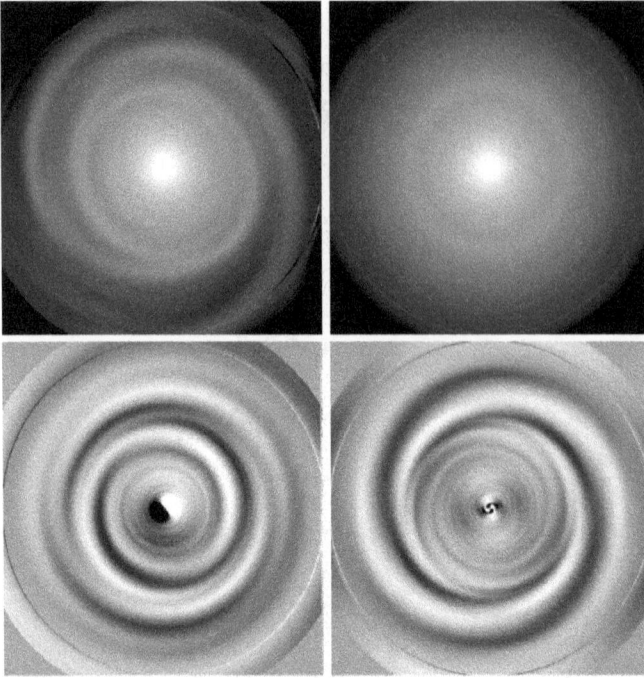

Figure 8 – Fourier decomposition of the I band image of NGC 4622. This is the stellar background light distribution. *Top left*: sum of m = 0-6 terms. *Top right*: m = 0 image. *Bottom left*: m = 1 image. *Bottom right*: m = 2 image. North is to the upper right, east to the upper left. Each frame covers a field of 1.50′ x 1.43′.

Answering The Common Question

Figure 8 shows the Fourier decomposition of the infrared image of NGC 4622 – this wavelength region reveals the stellar background light distribution. The m = 0 image shows the distribution for the underlying exponential disk. Stellar light typically follows an exponential distribution in a disk. The m = 1 image maps the single arm (one-fold symmetry). The m = 2 image maps the outer pair of arms (two-fold symmetry). Note that the m = 1 arm wraps around 540°. The inner arm is a clear component of the NGC 4622 structure.

Conclusion

So the galaxy that no one wants to see is clearly there with its unusual structure defying the accepted idea that all spirals have outer trailing arms. Byrd, Howard, Buta, and Freeman continue their work on this intriguing spiral investigating other methods to support their conclusion that the outer arms lead. Regardless, it is an unpopular idea. When this initial result was presented at a recent professional conference one of the attendees said "You are the backward astronomers who found a backward galaxy". Backward or not, NGC 4622 is clearly worth more study.

References

Binney, J. and Tremaine, S. 1987, *Galactic Dynamics*, Princeton, Princeton Univ. Press

Buta, R., Crocker, D., and Byrd, G. G. 1992, *AJ*, **103**, 1526

Buta, R., Byrd, G. G., and Freeman, T. 2003, *AJ*, **125**, 634

Byrd, G. G. et al. 1989, *Celestial Mech.*, **45**, 31

Byrd, G. G., Freeman, T., and Howard, S. 1993, *AJ*, **105**, 477

de Vaucouleurs, G. 1958, *ApJ*, **127**, 487

Hubble, E. 1943, *ApJ*, **97**, 112

Lindblad, B. 1941, Stockholms observatoriums annaler, bd. **13**, no 10, Stockholm
    Almqvist & Wiksell, 3

Introduction to

Black Holes Can Dance

Sethanne Howard is the retired Chief of the Nautical Almanac Office. The following article appeared in 2011 in the Journal of the Washington Academy of Sciences, Volume 97. Howard discusses black holes, those regions of space from which nothing, not even light, can escape. Briefly moving through the mathematics to illustrate the Schwarzschild and Kerr metrics, the article then discusses interesting facts about black holes, including a trip to a black hole, Hawking radiation, and frame dragging.

Black Holes Can Dance

Sethanne Howard

USNO, retired

Abstract

This is the story of black holes seen from one astronomer's perspective. Although some very technical information is included, the intent is to review some information about this odd piece of nature.

Introduction

SIMPLY PUT, A *BLACK HOLE*[i] is a region of space from which nothing, not even light, can escape. It is the result of the deformation of space-time caused by a *very* compact mass – a lot of mass in a teeny (actually zero) volume. Around the black hole there is an undetectable surface (called the event horizon) which marks the point of no return. Once inside nothing can escape. A black hole is called "black" because it absorbs all the light that hits it, reflecting nothing, just like a perfect blackbody in thermodynamics. We cannot see, hear, smell, touch, or taste it.

Now that we know that much, let us look at some history.

Newton's universe did not include black holes. I shall start there, and assume we know the basics of Newton's laws.

Even though Newton did not discuss black holes, the idea of them has been around for some time.

The idea of a body so massive that even light could not escape was first put forward by geologist John Michell[ii] in a letter written to Henry Cavendish[iii] in 1783 to the Royal Society:[iv]

> If the semi-diameter of a sphere of the same density as the Sun were to exceed that of the Sun in the proportion of 500 to 1, a body falling from an infinite height towards it would have acquired at its surface greater velocity than that of light, and consequently supposing light to be attracted by the same force in proportion to its *vis inertiae*, with other bodies, all light emitted from such a body would be made to return towards it by its own proper gravity.

In 1796, mathematician Pierre-Simon Laplace[v] promoted the same idea in the first and second editions of his book *Exposition du système du Monde* (it was removed from later editions). He pointed out that there could be massive stars whose gravity is so great that not even light could escape from their surface.

Such dark objects were ignored until the 20th century, since it was not understood how gravity could influence a massless wave such as light.

It took Albert Einstein (and others) to show that gravity can influence light. First with his Special Theory of Relativity and second with his General Theory of Relativity he proved that gravity does influence the motion of light. According to Einstein, space warps when close to matter. The more matter there is, the more space warps. The description of the curvature (warping) of space is the mathematically complicated part of general relativity. It involves tensor calculus and *metrics*. In mathematics, the word metric refers to a fairly general function which defines the 'distance' between elements in a set. I tend to think of a metric as a bendable and twistable ruler that allows one to measure intervals (distances) between two events. **Keep the concept of a metric in mind**.

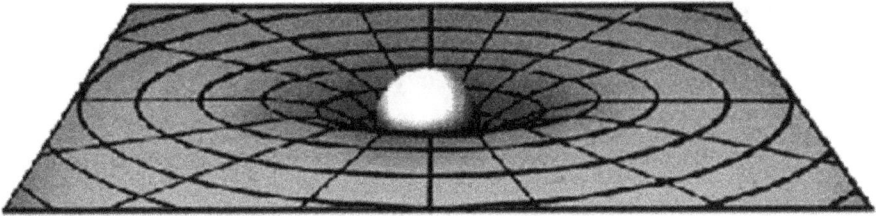

Figure 1. The curvature of space caused by a massive object.

Figure 1 represents a two-dimensional slice through three-dimensional space showing the curvature of space produced by a spherical object, *e.g.*, the Sun. Einstein's view is that the planets follow the curvature of space around the Sun (and produce a tiny amount of curvature themselves).

Metrics and the Special Theory of Relativity

The Special Theory of Relativity (STR) has as its basic premise that light moves at a uniform speed, $c = 300,000$ km/s, in all frames of reference. This results in setting the speed of light as the absolute speed limit in the universe and also produces the famous relationship between mass and energy, $E = mc^2$.

In Newtonian flat space (the kind we are familiar with) the metric that defines distance is:

$$ds^2 = dx^2 + dy^2 + dz^2$$

where ds is the distance and (x, y, z) are the spatial coordinates (remember your high school geometry). Strictly speaking this is the line element, not the metric. For the purpose here, I use the words interchangeably.

In STR the metric becomes a combination of time and space:

$$ds^2 = -c^2dt^2 + dx^2 + dy^2 + dz^2 .$$

In spherical coordinates it is:

$$ds^2 = -c^2dt^2 + dr^2 + r^2d\theta^2 + r^2 \sin^2 \theta d\phi^2$$

or more concisely

$$ds^2 = -c^2dt^2 + dr^2 + r^2d\Omega^2 .$$

It is from STR that we get the term *space-time* – space and time forming a single continuum, ds. Note the difference between this metric and the metric of Newton's world. In Einstein's world the distance (interval) between two events depends on the time *and* space intertwined.

The General Theory of Relativity

As useful as Newtonian mechanics may be, it is merely a limiting case of relativistic mechanics. The General Theory of Relativity (GTR) is the geometric theory of gravitation published by Albert Einstein in 1916. It is the current description of gravitation in modern physics.

We need GTR because **black holes require GTR for explanation**, yet GTR is a difficult subject no matter how one looks at it. This is what the basic equation of GTR looks like:

$$G_\mu^\nu + \Lambda g_\mu^\nu = \frac{8\pi G}{c^4} T_\mu^\nu$$

where G_μ^ν is the Einstein tensor,[vi] Λ is the cosmological constant,[vii] g_μ^ν is the metric tensor,[viii] and T_μ^ν is the stress-energy tensor. This equation describes the interaction of gravitation as a result of space-time being curved by matter and energy. The left side of the equation contains the information about how space is curved (the geometry), and the right side contains the information about the location and motion of the matter (the dynamics). When fully written out, the equations are a system of coupled, nonlinear, hyperbolic-elliptic partial differential equations.

You may now forget these equations because they are not necessary for the rest of the paper except to say that solutions to these equations under certain conditions give us black holes.

Two Metrics That Define Black Holes

Solutions to the Einstein's GTR equations are *metrics of space-time* – ways to describe gravity and mass interacting with each other. **The metric is the fundamental object of study for black holes**.

The first solution came in 1916 when astronomer Karl Schwarzschild (1873-1916) solved the equations for the particular case of a non-rotating spherically symmetric point mass.[ix,x] This point mass solution (where all the mass is concentrated into a single point) describes a black hole.

The metric solution for the point mass was named after Schwarzschild – the *Schwarzschild metric* defines the space-time environment near a black hole of mass *m*. The metric is spherically symmetric and non-rotating (no angular momentum). This is the simplest type of black hole. The metric only looks complicated:

$$c^2 d\tau^2 = \left(1 - \frac{2Gm}{c^2 r}\right) c^2 dt^2 - \left(1 - \frac{2Gm}{c^2 r}\right)^{-1} dr^2 - r^2 d\Omega^2 .$$

The quantity $\left(1 - \frac{2Gm}{c^2 r}\right)$ appears twice. It is there so that in the limit of large r and small m the metric reduces to the Newtonian gravitational field around a point mass. At $r = 0$ there is a true singularity.[xi] Note, however, the possibility of infinity when $r = 2Gm/c^2$. This particular value for r is called the *Schwarzschild radius* (r_s), a special radius that is quite useful, as we shall see.[xii]

It took some time for the next black hole solution to appear; however, in 1963, mathematician Roy Kerr found the exact solution for a rotating black hole. The more complicated *Kerr metric* for a black hole with angular momentum J is:

$$c^2 d\tau^2 = \left(1 - \frac{r_s r}{\rho^2}\right) c^2 dt^2 - \frac{\rho^2}{\Delta} dr^2 - \rho^2 d\theta^2 - \left(r^2 + \alpha^2 + \frac{r_s r \alpha^2}{\rho^2} \sin^2 \theta\right) \sin^2 \theta d\phi^2 + \frac{2r_s r \alpha \sin^2 \theta}{\rho^2} c dr d\phi ,$$

where r_s is the Schwarzschild radius, and the scale lengths α, ρ, and Δ are:

$$\alpha = \frac{J}{mc},$$

$$\rho^2 = r^2 + \alpha^2 \cos^2 \theta, \text{ and}$$

$$\Delta = r^2 - r_s r + \alpha^2.$$

At $r = 0$ there is the true singularity, however, the Kerr metric has two values for r where it appears to be singular: r_{inner} and r_{outer}. The inner surface occurs where the purely radial component of the metric goes to infinity:

$$r_{inner} = \frac{r_s + \sqrt{r_s^2 - 4\alpha^2}}{2}.$$

The other singularity occurs where the purely temporal component of the metric changes sign from positive to negative:

$$r_{outer} = \frac{r_s + \sqrt{r_s^2 - 4\alpha^2 \cos^2 \theta}}{2}.$$

The Kerr black hole, therefore, has two special radii with the *ergosphere* (sphere of influence of the black hole – more on it later) between them. The outer surface is also called the static limit. The inner surface is also called the event horizon.

Note something important. The parameter t (time) does not occur in the right side of either metric. **Time stops at a black hole**.

So by the 1960s scientists could describe the enivironment around stationary,[xiii] non-rotating, and rotating black holes.[xiv] Given these metrics, people got to work on the dynamics of black holes.

The Four Laws

By the 1970s research by many people led to the formation of the four laws of black hole dynamics. These laws describe the behavior of a black hole in close analogy to the laws of thermodynamics by relating mass to energy, surface area to entropy, and surface gravity to temperature. The analogy was completed when Stephen Hawking, in 1973, showed that quantum field theory predicts that black holes radiate (*Hawking radiation*, see the section on this) like a blackbody with a temperature proportional to the surface gravity of the black hole.[xv] Further description of the four laws is highly mathematical and beyond the scope of this paper.

Despite these laws we still cannot describe a black hole all the way to $r = 0$. That will require combining quantum and gravitational effects into a single theory, although the single theory does have a name: quantum gravity. This is an area of active research.

However, the four laws led to a definition of what one *can* measure in a black hole.

Black Holes Have No Hair

The four laws led to the 'no-hair theorem' – black holes have no hair. This means that a stationary black hole is completely described by only three things: its *mass, angular momentum,* and *electric charge*. There is no other way to 'grab' onto (measure) a black hole. These properties are special because they and only they are detectable from outside the black hole. For example, a charged black hole repels other like charges just like any other charged object. Why these three? The reason is mathematical; these are unique, conserved imprints in the external fields of the black hole (conserved Gaussian flux intervals).

Theoretically a black hole may possess electric charge but it would quickly attract charge of the opposite sign and become neutral. The net result is that any realistic black hole would tend to exhibit zero charge.

I shall discuss two types of black holes: one defined by the Schwarzschild metric and one defined by the Kerr metric. There are others, but they are quite specialized. In fact, there are four basic types:

	Non-rotating	Rotating
No charge	Schwarzschild	Kerr
Charged	Reissner-Nordström	Kerr-Newman

When most people think of a black hole, it is usually the Schwarzschild black hole. So I shall discuss this one first.

The Particulars of the Schwarzschild Black Hole

The Schwarzschild black hole is stationary (not moving through space) with zero charge and is non-rotating. It is 'dead' in the sense that one can never extract from it any of its mass-energy. No information can ever come from a Schwarzschild black hole. This means it is stable against a perturbation (*e.g.*, a kick), were you so inclined.

At the Schwarzschild radius, r_s, some of the terms in the metric apparently become infinite. This is not a true singularity.[xvi] It is due to the choice of spherical coordinates; however, it does have a physical effect.

In 1958 physicist David Finkelstein[xvii] identified the Schwarzschild surface $r_s = 2Gm/c^2$ as an *event horizon,* "a perfect unidirectional membrane: causal influences can cross it in only one direction." That means it is a one way gate. Things go in and do not come out. At the rim of the event horizon one must travel at the speed of light just to stay in place. Once inside the event horizon the radial coordinate 'evaporates' because there can be *no* spatial direction that will lead back to the outside. **Once inside the event horizon escape is not possible**.

The 'size' of a black hole, as determined by the radius of its event horizon (Schwarzschild radius), is roughly proportional to its mass M:

$$r_s = \frac{2GM}{c^2} \approx 2.95 \frac{M}{M_\odot} \, \text{km} ,$$

where $M\odot$ is the mass of the Sun. This relation is exact only for Schwarzschild black holes; for more general black holes it can differ up to a factor of two. Table I shows this relation for some common objects.

Table I. The Schwarzschild radius for different sized objects.

Object	$r_{\text{Schwarzschild}}$
Earth	1 cm
Jupiter	3 meters
Sun	3 kilometers
3 solar-mass star	9 kilometers
3000 solar masses	9000 kilometers

So, for the Earth to become a black hole, all its mass must be consolidated within a sphere of a one centimeter radius. This is highly unlikely.

Figure 2 illustrates how space-time curves about a Schwarzschild black hole. At the center of a black hole lies a true gravitational singularity.[xviii] At $r = 0$ the space-time curvature becomes infinite. For a non-rotating black hole this region takes the shape of a single point. For a rotating (Kerr) black hole it is smeared out to form a ring singularity lying in the plane of rotation. In both cases the singular region has zero volume. The singular region can thus be thought of as having infinite density. All this really means is that we do not understand what happens at the singularity.

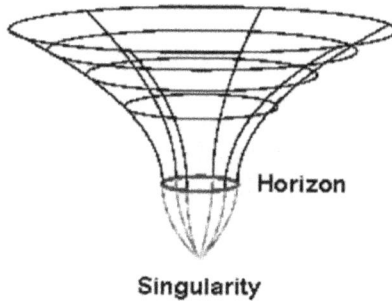

Horizon

Singularity

Figure 2. Curved space-time around a Schwarzschild black hole.

Does this mean that gravity is somehow different around a black hole? It is misleading to say that black holes have 'stronger' gravity than other masses. Black hole or not, the curvature one feels depends strictly on the mass of the object and the distance one is from that mass, not whether the object is a black hole. When that mass is concentrated in a small volume, however, one can get closer to the mass than otherwise is the case. This may be why one thinks gravity is stronger around a black hole. It is actually the field density that is greater. The gravitational field density close to an Earth mass compressed within a 1 cm radius is much higher than the density around an Earth mass with its current radius. A lot of mass in a very tiny space can strongly warp the space nearby – tiny and nearby being the key words.

So, if the Sun became a black hole, would we on Earth notice? We would miss the sunlight and die (so we would notice), but the gravitational effect on Earth would be what it always was. The mass of the Sun has not changed (even though it occupies a smaller volume); the Earth's distance from the Sun has not changed; therefore, the Earth feels the same effect and continues to orbit the black hole Sun.

This leads me to the next point.

Black holes are *not* cosmic vacuum cleaners. They do not zoom around space sucking up matter. The black hole Sun will not scoop up the Earth. Far from the Sun there is no unusual gravitational influence. Only within a few Schwarzschild radii is there a significant effect. Black holes *can* accrete matter but only when the matter is quite close. In fact, scientists believe black holes are surrounded by *accretion disks* – a disk of accreting matter (usually gaseous) orbiting the central object.

Now that we know a bit more about them, let us see how they are made.

Making a Black Hole

There are two classes of black holes: (1) a stellar mass black hole, and (2) a super massive black hole.

Class (1) happens when a heavyweight star reaches the end of its life. One way to classify stars is by their birth weight. Heavyweight stars are born with more than 30 solar masses to their credit. Most of a star's life is spent maintaining a balance between two forces: radiation pressure from the nuclear fusion in the core that pushes outward; and the gravitational force trying to compress the gas inward. Ultimately a heavyweight star will, as all stars do, consume all its nuclear fuel. It can then no longer support itself against a subsequent gravitational collapse. If it fails to eject its excess mass in the collapse process then nothing can stop the stellar remnant from collapsing toward a point – forming a black hole. This collapse happens in milliseconds – the star winks out. Our Sun (a lightweight star) is not massive enough to end its life this way. It will end as a white dwarf. A middleweight star (between 6 and 30 solar masses) will explode as a supernova and end as a neutron star.

Most stars are not perfectly spherical and have a lot of angular momentum, so the gravitational collapse produces a black hole more in line with a Kerr black hole than a Schwarzschild black hole.

Class (2) is thought to lurk in the centers of most galaxies. A super massive black hole contains thousands to millions of solar masses. Once a super massive black hole has formed, it can continue to grow by absorbing additional matter. One model for the formation of super massive black holes is by slow accretion of matter onto a stellar mass black hole. Another model involves a large gas cloud collapsing into a relativistic star of perhaps a hundred thousand solar masses or larger. The star would then become unstable and may collapse directly into a black hole without a supernova explosion. Super massive black holes have properties which distinguish them from stellar mass black holes:

- The average density of a super massive black hole (defined as the mass divided by the volume within its Schwarzschild radius) can be as low as the density of water for very high mass black holes.
- The tidal forces in the vicinity of the event horizon are significantly weaker than the tidal forces around a stellar mass black hole.

Astronomers think the universe is littered with black holes, that they are not rare at all. In addition to the stellar type they think that nearly every galaxy has a central super massive black hole. What if we could visit one?

A Trip to a Black Hole

An observer falling into a Schwarzschild black hole cannot avoid the singularity at $r = 0$. Any attempt to do so will only shorten the time taken to get there. As the traveler spirals in, there is a last stable orbit[xix] at a distance of $3r_s$. Continuing inward, the traveler crosses the event horizon. At the singularity the traveler is crushed to infinite density and its mass added to the black hole. Before that happens, though, it will have been torn apart by the tidal forces in a process sometimes referred to as *spaghettification* (I did not make that up) or the *tube-of-toothpaste-effect*.

To describe this in more detail, assume there are two astronauts, a smart one and a dumb one. Their spaceship arrives near a 3 solar mass black hole ($r_s = 9$ km). The smart astronaut stays in the spaceship. The dumb astronaut jumps toward the black hole. Let us pick up the action 900 km away ($100r_s$).

At this distance of 100 r_s from the black hole the dumb astronaut is torn apart due to the tidal effect of gravity, and the story ends. For comparison, the gravity tide on a human (head to toe) on the Earth's surface is about one millionth of a g.

For a 3 solar mass black hole the tidal force of the black hole is shown in Table II. Remember that tidal forces go as $1/r^3$, so the tides quickly become fatal as one approaches the black hole.

Table II. Tidal force in g's versus distance in km from a 3 solar mass black hole.

Distance (km)	Force (g's)
6400	0.5
2000	18
1000	144
100	150,000
10	150,000,000

For the purposes of argument, though, assume the dumb astronaut is 'stretchable.' Then as he falls, toe first, his toes are closer to the upcoming event horizon than his head. The gravity tides between his toes and head cause his toes to travel faster than his head. He stretches. The closer he gets, the more he stretches. Simultaneously he is squeezed into regions of ever decreasing circumference. He gets longer and thinner, forming dumb astronaut spaghetti strings.

As he travels toward the event horizon he may notice nothing out of the ordinary, except an inability to steer himself in any but one direction – which is toward the "invisible" hole. He will never know when he has crossed the event horizon were it not for the increased tidal tugging that draws his body longer and longer, squeezing in from the sides (actually at this point he is a set of disconnected atoms, zooming along all in a line). Just before he reaches the event horizon, each piece of him emits high energy radiation (x-rays) as that piece disappears forever. He winks out of sight with a puff of radiation. It is a rather spectacular way to die.

And it is a wonderful way to garner energy. The efficiency of energy generation near a stationary black hole is about 6%. Near a rotating black hole this reaches about 30% efficiency. This is a staggering amount. It is the best return of energy known. Compare the efficiency of combustion on Earth which is only about 10^{-8}. The efficiency of nuclear burning (in a star) is about 7×10^{-3}.

A visit to a super massive black hole is less dramatic although it ends the same way. If the mass of the black hole is about 30,000 solar masses then the dumb astronaut will not be torn apart by tidal forces at the event horizon. This will wait until he is much deeper inside. Of course, once he crosses the event horizon he cannot return or send messages. Although he may survive those tidal forces, the high energy radiation (all those x-rays and gamma rays lurking at the event horizon) will fry him.

One can calculate how long the dumb astronaut spaghetti string will "live" once inside the event horizon. No matter how he approaches a black hole of mass M, once inside the event horizon he will be killed at the $r = 0$ singularity in a proper time of about 1.54×10^{-5} $M/M_\odot$ seconds. So for a 10 solar mass black hole, he will die in 154×10^{-5} sec (0.00154 sec). One way or another, **the dumb astronaut will not survive the trip**.

Time Dilation

If the dumb astronaut carries a flashlight and points it back at the smart astronaut, and flashes it in a regular pattern, what will the smart astronaut see? She will see the flashes get further and further apart eventually slowing down to a stop (after an infinite amount of time). The GTR predicts that time will slow in the presence of matter – this is called time dilation. It is not just clocks by the way, all physical processes, including clocks ticking (however they measure their ticks), hearts beating, aging, *etc.*, must slow down, but the only one who notices is the distant timekeeper. This is not an imaginary effect. When transporting atomic clocks on the Earth, one must correct for the GTR effects of the Earth on the moving clock.

Gravitational Redshift

In addition to the slow down of time, the light she sees is redshifted more and more as the dumb astronaut gets closer to the event horizon. This is not a Doppler shift. Light loses energy when escaping from a gravitational field. Because the energy of light is proportional to its frequency, a shift toward lower energy represents a shift toward the red for visible light. This gravitational redshift was first observed in the spectra of dense white dwarf stars, whose light is redshifted by about 1Å. Gravitational redshift was experimentally verified on Earth by the Pound–Rebka experiment of 1959.

Now that we know the dumb astronaut will not survive, is there a way we can we tell that he fell in?

Cosmic Censorship

When an object falls into a black hole, any and all information about the shape of the object or distribution of charge on it is evenly distributed along the horizon of the black hole, and is lost to outside observers. So not only does the dumb astronaut disintegrate, but also there is no way to determine that it was a dumb astronaut that fell in.

Because the black hole eventually achieves a stable state with only three measureable parameters (mass, charge, and angular momentum), there is no way to avoid losing information about the initial conditions. Nature puts a curtain around black holes so that we cannot see inside or know what happens inside – this cosmic censorship is complete. **There are no 'naked' singularities**. And the event horizon must be real, not complex.[xx] The information that is lost includes every quantity, including the total baryon number, lepton number, and all the other nearly conserved pseudo-charges of particle physics.

At least that was the state of the current thought until the 1970s. The physicist Stephen Hawking (author of *A Brief History of Time* and *The Universe in a Nutshell*) has long worked in theoretical cosmology. In 2009 he received the Presidential Medal of Freedom. He even played himself on *Star Trek*. I discuss one aspect of his work next.

Hawking Radiation

In 1974 Stephen Hawking realized that black holes are not absolutely black. There are quantum effects that allow black holes to emit blackbody radiation. The temperature of this radiation is inversely proportional to the black hole's mass; the tinier the black hole the higher the temperature of the radiation – called *Hawking radiation.*

Hawking radiation is due to particle/anti-particle pairs (*e.g.*, electron/positron) which are continuously created and annihilated in free space. When this pair creation happens near a black hole it is possible for one of the two particles to cross the event horizon before it meets and annihilates its partner. The other particle is then free to leave the scene, making the black hole appear to the outside world as a source of radiation. In other words, there is 'new' energy. So to satisfy energy conservation, the particle that fell in must have a negative energy (with respect to an observer far away from the black hole). Thus, the black hole loses mass, and, to an outside observer, it appears that the black hole has just emitted a particle. It takes energy to create new particles. This energy must come from the black hole. The black hole therefore decreases its mass as it radiates. Thus black holes will slowly evaporate.

A one solar mass black hole has a temperature of only 60 nanoKelvin (v-e-r-y cold); in fact, such a black hole would absorb far more cosmic microwave background radiation than it emits. Evaporation will take 10^{70} years. This is far longer than the age of the universe.

A smaller black hole of 4.5×10^{22} kg (about the mass of the Moon) would be in equilibrium at 2.7 K, absorbing as much radiation as it emits. Even smaller black holes would emit more than they absorb, and thereby lose mass.

For a miniature black hole – about 10^{12} kg mass which is the mass of a mountain – evaporation will take about as long as the universe is old. It is conceivable that conditions in the very earliest epochs of the universe might have been just right to compress pockets of matter into these miniature black holes. The Schwarzschild radius of such a black hole is about 10^{-15} m, comparable to the size of a subatomic particle. This begs the question how these teeny black holes got formed; however, at the end of its life, the mass of the tiny black hole becomes smaller and smaller, and hence its temperature tends towards infinity. The black hole ultimately disappears in an explosion. Fortunately (or unfortunately) current physics is unable to explain the last phases of the evaporation of the black hole.

Black Hole in a Bathtub

Recently scientists in Canada have measured the equivalent of Hawking radiation from a "bathtub black hole." In August 2010, the Canadian scientists announced[xxi] that they had made an event horizon in a water channel. They sent a steady flow of water in one direction. As it passed over the top of a piece of wood whittled in the shape of an airplane wing, the water traveled faster (aka Bernoulli). In the opposite direction, the group created water waves. When these waves approached the wing, where water was flowing faster, they slowed to a stop. Technically this bathtub version is a white hole, an inverted black hole that keeps waves out rather than bringing them in. But the white hole serves as an analog because it shares an important feature with astrophysical black holes — an imaginary boundary that emits an unusual kind of radiation. These laboratory emitters of Hawking type radiation share one required feature with their astrophysical counterparts — a point of no return, analogous to the black hole's event horizon. Both types have event horizons, so both ought to emit Hawking radiation. In fact, pairs of short-wavelength waves were created at the bathtub horizon and swept away, and the energy of these emitted waves matches what would be predicted from Hawking radiation around a real black hole.

This seems a bit forced to me.

Hawking radiation introduced a debate in cosmological circles. Is it consistent with the no hair theorem? This leads to a paradox.

The Paradox in Hawking Radiation

There is a paradox with Hawking radiation. From the no hair theorem, one expects the Hawking radiation to be completely independent of the material entering the black hole. All information is lost entering a black hole. Nevertheless, if the material entering the black hole were a pure quantum state, the transformation of that state into the mixed state of Hawking radiation would destroy information about the original quantum state. The rules of quantum mechanics say information is conserved in the wave function. The no hair theorem says the information is lost – a physical paradox.

Hawking remained convinced that the equations of black hole thermodynamics together with the *no-hair theorem* led to the conclusion that quantum information will be destroyed. This annoyed many physicists, notably John Preskill, who in 1997 bet Hawking and Kip Thorne that information was not lost in black holes. This led to the

Susskind-Hawking battle, where Leonard Susskind and Gerard't Hooft publicly declared war on Hawking's solution, with Susskind publishing a popular book about the debate in 2008 (*The Black Hole War: My battle with Stephen Hawking to make the world safe for quantum mechanics*). The book carefully notes that the war was purely a scientific one, and that at a personal level, the participants remained friends. The solution to the problem is the holographic principle (a property of quantum gravity combined with string theory). With this, as the title of an article puts it, "Susskind quashes Hawking in quarrel over quantum quandary."

In July 2005, Stephen Hawking announced a theory that quantum perturbations of the event horizon could allow information to escape from a black hole, which would resolve the information paradox. When announcing his result, Hawking also conceded the 1997 bet, paying Preskill with a baseball encyclopedia "from which information can be retrieved at will." However, Thorne remains unconvinced of Hawking's proof and declined to contribute to the award.

It does not end there. Roger Penrose advocates the Conformal Cyclic Cosmology (CCC) which critically depends on the condition that information is in fact lost in black holes. In CCC, the universe iterates through infinite cycles, with the future time-like infinity of each previous iteration being identified with the Big Bang singularity of the next. This CCC model might in future be tested experimentally by detailed analysis of the cosmic microwave background radiation (CMB): if true the CMB should exhibit circular patterns with slightly lower or slightly higher temperatures. In November 2010, R. Penrose and V. G. Gurzadyan announced they had found evidence of such circular patterns (Figure 3), in data from the Wilkinson Microwave Anisotropy Probe corroborated by data from the BOOMERanG experiment[xxii]. However, the statistical significance of the claimed detection has been questioned. Three groups have independently attempted to reproduce these results, and found that the detection of the concentric anomalies was not statistically significant. Stay tuned.

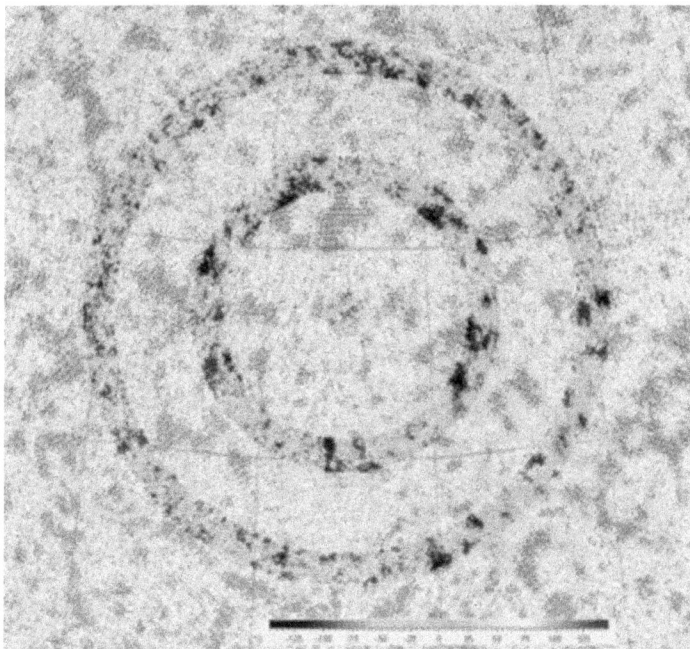

Figure 3. Possible circles in BOOMERanG data – the concentric circles are highlighted. The mottling represents the wrinkles in space-time of the cosmic microwave background.

The Particulars of the Kerr Black Hole

The other type of black hole I shall discuss is the Kerr black hole, which rotates (has angular momentum). Seen in cross-section, the Kerr black hole is oval-shaped, with the ergosphere extending farther into space at the black hole's equator than at its poles (Figure 4). The $r = 0$ singularity is a ring of zero volume.

The Kerr black hole is actually more significant than the Schwarzschild black hole because most black holes spin. Part of the mass is actually stored as rotational energy in the ergosphere (which means 'place where work can be done') and is, in theory, available for extraction since the mass has not yet crossed the event horizon. This type can inject energy into its surroundings – hence this type is 'live.'

Kerr space-time is what happens when a black hole has reached its final evolutionary state. Kerr space-time is time-independent, meaning that nothing in Kerr space-time changes over time. In effect, time stands still. Remember that the time parameter t does not appear in the right side of the Kerr metric. A black hole in such a state is essentially stationary.

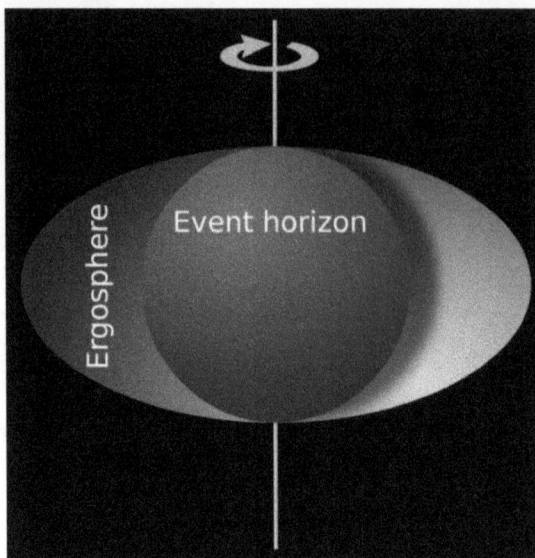

Figure 4. Side view of a conceptual Kerr black hole.

Frame Dragging

When the black hole is spinning it actually pulls the fabric of space-time around with it – an effect called frame dragging, also known as the *Lense-Thirring* effect. The rotation of the black hole or even the rotation of a very massive object will alter local space-time by dragging a nearby object out of position compared with the predictions of Newtonian physics. Frame dragging is like what happens if a bowling ball spins in a thick fluid such as molasses. As the ball spins, it pulls the molasses around itself. Anything stuck in the molasses will also move around the ball. This dragging happens in the ergosphere. The closer to the black hole the greater the dragging.

Inside the ergosphere (inside the static limit) nothing can stand still; therefore, particles falling within the ergosphere are forced to rotate and thereby gain energy. They *must* orbit in the same direction as the black hole rotates. So long as they are still outside the event horizon, they may, however, escape the black hole. The net process is that the rotating black hole emits energetic particles at the cost of its own total energy. The possibility of extracting spin energy from a rotating black hole was first proposed by the mathematician Roger Penrose in 1969.

The Earth is a very massive object; therefore, as the Earth rotates, it pulls space-time in its vicinity around itself. This action introduces a precession on all gyroscopes in a stationary system surrounding the Earth (Figure 5). The predicted Lense-Thirring effect is small — about one part in a few trillion – yet measureable. A Foucault pendulum would have to oscillate for more than 16000 years to precess 1 degree.

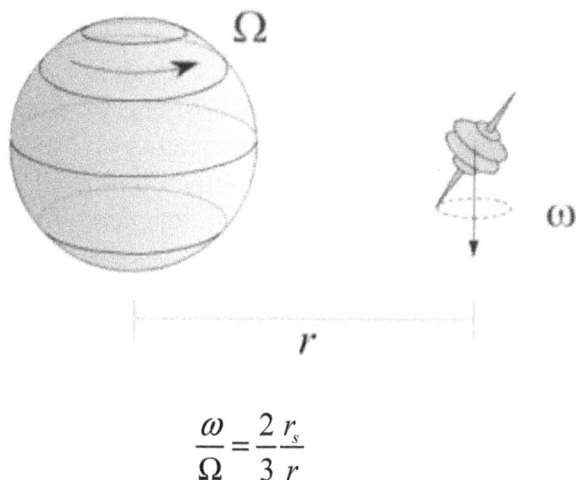

$$\frac{\omega}{\Omega} = \frac{2}{3}\frac{r_s}{r}$$

Figure 5. The Earth rotating with angular velocity Ω; The gyroscope a distance r away precesses with angular velocity ω.

LAGEOS (Laser Geodynamics Satellites) are a series of scientific research satellites designed to provide an orbiting laser ranging benchmark for geodynamical studies of the Earth. The Lense-Thirring effect on LAGEOS due to the rotating Earth has been measured. The effect shifts the orbits of the satellites about 2 meters per year in the direction of rotation. The results are compatible with the predictions of GTR.

Another test of frame dragging is the Gravity Probe B satellite – launched in 2004 (now decommissioned) with a dual purpose: to measure the frame dragging of Earth, and to measure the geodetic effect – the amount by which the Earth warps the local space-time in which it resides. For Gravity Probe B, in polar orbit 642 km above the Earth, frame dragging causes its gyroscope spin axes to precess in the east-west direction by a mere 39 milliarcsec/yr. — an angle so tiny that it is equivalent to the average angular width of Pluto as seen from Earth.

Initial results from Gravity Probe B confirmed the expected geodetic effect to an accuracy of about 1%. In December 2008 NASA reported that the geodetic effect was confirmed to better than 0.5%. Unfortunately the expected frame-dragging effect was similar in magnitude to the noise level. Work continues on the data to model and account for these sources of unintended signal, thus permitting extraction of the frame-dragging signal if it exists at the expected level. By August 2008 the uncertainty in the frame-dragging signal had been reduced to 15%. Final results are expected in 2011.

Many astrophysical objects, *e.g.* pulsars and black holes, emit jets of energy. These jets may also provide evidence for frame-dragging. Such jets are extremely powerful bursts of energy. Some of them extend huge distances into space. There are images of jets later in the paper (Figures 12, 13). These jets are tightly collimated flows of energy, collimated perhaps by the twisting of magnetic field lines by frame dragging.

The energy released in an astrophysical jet is overwhelmingly powerful – at the highest end of the electromagnetic spectrum – x-rays and gamma rays. A trip through a jet would quickly fry the traveler. Or is there a way to cut through space-time?

Shortcuts through Space – Wormholes?

If one can avoid the jet, perhaps one can escape to another universe. A wormhole is a hypothetical topological feature of space-time that would be, fundamentally, a "shortcut" through space-time. The physicist John Wheeler coined the term *wormhole* in 1957; however, in 1921, the mathematician Hermann Weyl already had proposed the wormhole theory.

There is no observational evidence for wormholes, but there are valid theoretical solutions to the equations of GTR which contain wormholes. These solutions say that it is theoretically possible to avoid the singularity at $r = 0$ and exit the black hole into a different space-time with the black hole acting as a wormhole.

For a simple explanation of a wormhole, consider space-time as a two-dimensional (2D) surface. If this surface is folded along a third dimension, it allows one to picture a wormhole "bridge." (Please note that this is merely a visualization to convey a structure existing in four or more dimensions). The parts of the wormhole could be higher-dimensional analogues for the parts of the curved 2D surface; for example, instead of mouths which are circular holes in a 2D plane, a real wormhole's mouths could be spheres in 3D space. A wormhole is, in theory, much like a tunnel with two ends each at separate points in space-time. Figure 6 illustrates a 2D wormhole.

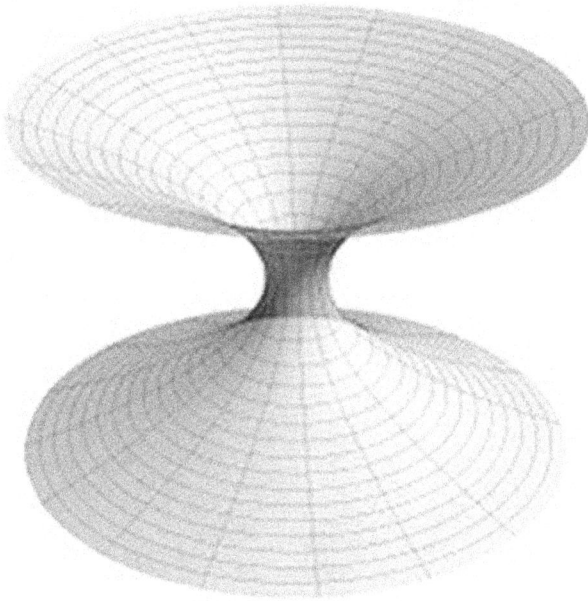

Figure 6. A 2D representation of a wormhole.

The first type of wormhole solution discovered was the *Schwarzschild wormhole*. Technically the Schwarzschild metric has a negative square root as well as a positive square root solution for the geometry. The complete Schwarzschild geometry consists of a black hole, a white hole, and two universes connected at their event horizons by a wormhole. The negative square root solution inside the horizon represents a white hole. A white hole is a black hole running backwards in time. Just as black holes swallow things irretrievably, so also do white holes spit them out.[xxiii] The negative square root solution outside the event horizon represents another universe. The wormhole joining the two separate universes is known as the Einstein-Rosen bridge. Unfortunately it is impossible for a traveller to pass through this wormhole from one universe into the other. A traveller can pass through an event horizon only in one direction. First, the traveller must wait until the two white holes have merged, and their horizons meet. The traveller may then enter through one horizon. But having entered, the traveller cannot exit, either through that horizon or through the horizon on the other side. The fate of the traveller who ventures in is to die at the singularity which forms from the collapse of the wormhole.

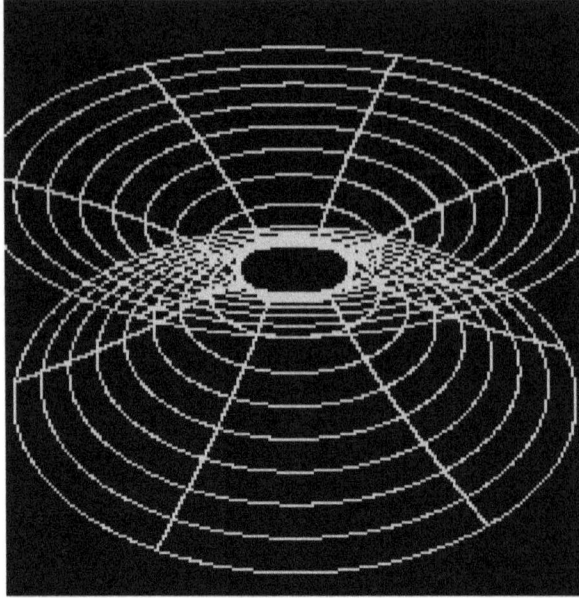

Figure 7. A 2D representation of a traversable wormhole.

Wormholes which could actually be crossed (Figure 7), known as *traversable wormholes*, would only be possible if exotic matter with negative energy density could be used to stabilize them (keep them from collapsing). Physicists have not found any natural process which would form a stable wormhole, although the quantum foam hypothesis (QFH) is sometimes used to suggest that tiny wormholes might appear and disappear spontaneously at the tiniest scale. Qualitatively QFH is described as subatomic space-time turbulence at extremely small distances of the order 10^{-35} meters. At such small scales of time and space the Heisenberg uncertainty principle allows particles and energy to come briefly into existence, and then annihilate, without violating conservation laws. However, without a theory of quantum gravity it is impossible to be certain what space-time would look like at these scales.

Finally, even if wormholes exist and are stable, they are quite unpleasant to travel through. Radiation that pours into the wormhole (from nearby stars, the cosmic microwave background, jets, *etc*.) gets blueshifted to very high frequencies, well into the high energy end of the spectrum. As you try to pass through the wormhole, you will get fried by these x-rays and gamma rays. So, at the moment, **space travel using wormholes is not possible.**

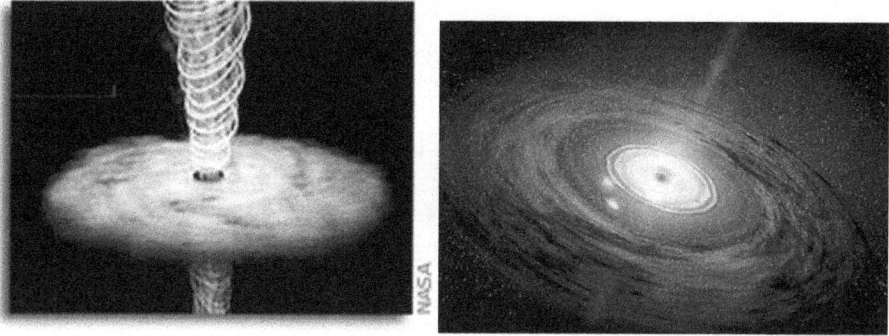

Figure 8. Two artists' conception of a black hole surrounded by an accretion disk and bipolar jets.

Observational Evidence for Black Holes

All this fancy theory means little unless there is observational evidence. Fortunately, despite its invisible interior, a black hole can be detected through its interaction with other matter.

There are several types of interactions. Five of them are: (1) A black hole exerts gravitational pull on surrounding matter, although this is indistinguishable at $r \gg r_s$, from the pull of an object with the same mass; (2) Gas surrounding a black hole is pulled inward and heated so that it emits x-rays and gamma rays that might be observed; (3) A lump of matter falling into a black hole should emit a burst of gravitational waves; (4) Tidal forces will tear matter apart and eject a blob of relativistic matter (tube-of-toothpaste effect); (5) Frame dragging will twist the magnetic field lines that may surround a black hole and thus 'shake' the external plasma.

Figure 8 gives two artists' conception of a black hole surrounded by an accretion disk with two polar jets. Conservation of angular momentum means gas falling into the gravitational well created by a massive object will typically spiral in to form a Frisbee-like structure (accretion disk) around the object. **Accretion disks are where the action is.** In the case of black holes, the accretion disk is outside the event horizon. The gas in the inner regions (closer to the event horizon) becomes so hot that it will emit vast amounts of radiation (mainly x-rays), which may be detected by telescopes. In many cases, accretion discs are accompanied by relativistic jets emitted along the poles, which carry away much of the energy. The mechanism for the creation of these jets currently is not well understood, although frame dragging is part of the solution.

It is unlikely we can observe an accretion disk directly (they are too small), but the jets are easily seen (there are examples later in the paper).

The strongest evidence for black holes comes from binary star systems in which a visible star orbits a massive but unseen companion. Binary x-ray sources[xxiv] are excellent candidates for black holes because matter from the accretion disk streaming into the black hole is ionized and greatly accelerated, producing x-rays.

In 1972 an x-ray source (named Cygnus X-1) was discovered in the constellation Cygnus. The Cyg X-1 system has a blue supergiant star (HDE226868), about 25 times the mass of the sun, orbiting the x-ray source. So something non-luminous is there (neutron star or black hole). Figure 9 is an artist's conception of the Cyg X-1 system. The indirect evidence for the black hole Cyg X-1 is a good example of the search for black holes.

Doppler studies of the blue supergiant indicate a revolution period of 5.6 days about the dark object. Using that period plus spectral measurements of the visible companion's orbital speed leads to a calculated total system mass of about 35 solar masses.[xxv] The calculated mass of the dark object then is 8 to 10 solar masses; much too massive to be a neutron star which has an upper limit of about 3 solar masses – hence black hole. Figure 10 is an image of the system. The jet is clearly seen.

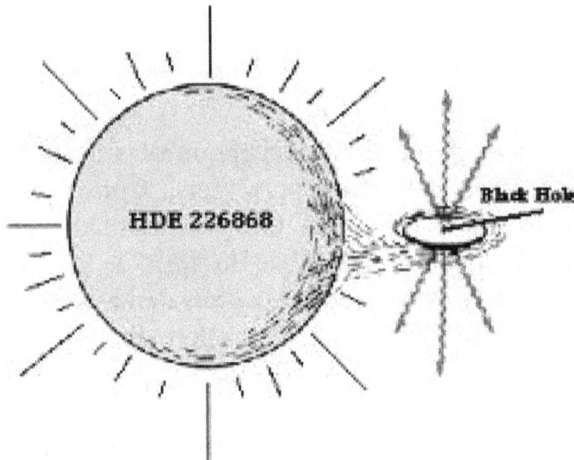

Figure 9. Artist's conception of Cygnus X-1 – matter is drawn from the supergiant star into an accretion disk around the black hole.

Further evidence for a black hole is the emission of x-rays from its location, an indication of temperatures in the millions of degrees. This x-ray source exhibits rapid variations, with time scales on the order of a millisecond. The light travel time is then a light-millisecond. This suggests a source not larger than a light-millisecond (300 km), so it is very compact. The only possibility that would place that much matter in such a small volume is a black hole.

Figure 10. Jet in Cyg X-1, the jet is coming out from the center toward 1 o'clock. The accretion disk is much too small to see.

In November 2010 evidence of the youngest black hole (all of 30 years old) known to exist in our cosmic neighborhood was found. This object provides a unique opportunity to watch a black hole develop from near infancy. The object is a remnant of supernova 1979C, a supernova in the galaxy M100 approximately 50 million light years from Earth. The scientists think the progenitor star for the supernova was a star about 20 times more massive than the Sun.

Astronomers have identified numerous stellar black hole candidates, and have also found evidence of super massive black holes at the center of galaxies. In 1998, astronomers found compelling evidence that a super massive black hole is located near the Sagittarius A* region (a bright and very compact astronomical radio source discovered in 1974 at the center of our own Milky Way). Astronomers monitored the orbits of individual stars very near the black hole and used Kepler's laws to infer the enclosed mass. Recent results indicate that the super massive black hole is 4.31 ± 0.38 million solar masses. Ultimately, what is seen is not the black hole itself, but observations that are consistent only if there is a black hole present near Sgr A.* There is a nice time lapse movie of the stellar motions in the area: http://apod.nasa.gov/apod/ap001220.html.

Super massive black holes can produce amazing jets. Figure 11 shows three jets from the object 3C 75 (object number 75 in the Third Cambridge Catalogue of radio sources).

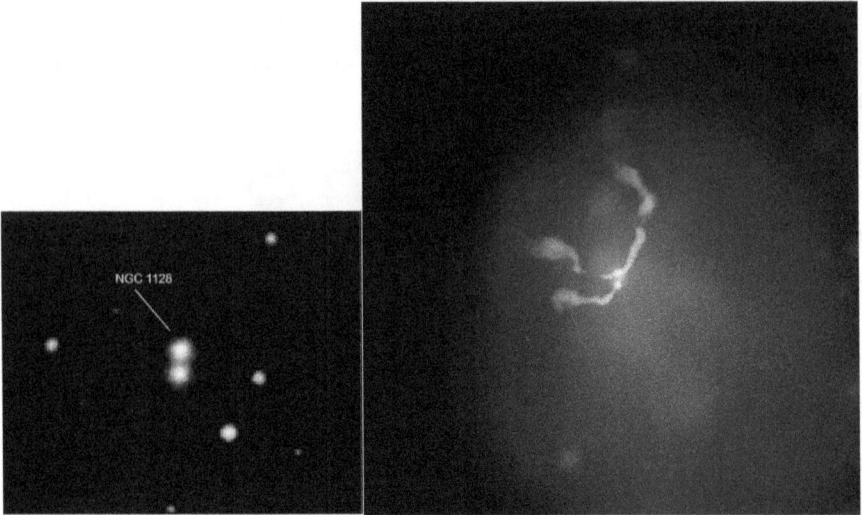

Figure 11. The right image is of 3C 75 – there are three clear jets. They originate at the bright spot in the center. The jets flare and bend as they encounter the intergalactic medium. The left image is an optical image of the galaxy NGC 1128 – the central bright dots in the right image.

The jets emanate from the vicinity of two super massive black holes (coming from the bright spot in the right image). These black holes are in the dumbbell galaxy NGC 1128, which has produced the giant radio source, 3C 75. The jets can reach incredible lengths – megaparsecs[xxvi] – streaming into intergalactic space.

The peculiar dumbbell structure of this galaxy is thought to be due to two large galaxies that are in the process of merging. Such mergers are common in the relatively congested environment of galaxy clusters. An alternative hypothesis is that the apparent structure is the result of a coincidence in time when the two galaxies are passing one another, like ships in the cosmic sea.

There is more. Black holes can come in pairs! Galaxies commonly collide and merge to form new, more massive galaxies. A merger between two galaxies should bring two super massive black holes to the new, more massive galaxy formed from the merger. The two black holes gradually spiral towards the center of this new galaxy, engaging in a gravitational tug-of-war with the surrounding stars. The result is a black hole dance.

142

Astronomers expect many such waltzing super massive black holes in the universe, but until recently only a handful had been found. In January of 2010, astronomers announced the discovery of 33 pairs of waltzing black holes in galaxies. This result shows that super massive black hole pairs are more common than previously known from observations. Also, the black hole pairs can be used to estimate how often galaxies merge with each other.

The largest known black hole inhabits the core of M87, a giant elliptical galaxy in the constellation Virgo. The M87 black hole appears to be about $(6.4 \pm 0.5) \times 10^9$ solar masses, with an event horizon diameter of about 18 billion km – almost twice the diameter of the orbit of Pluto. Figure 12 contains a series of photos of M87 with its jet. Surrounding the black hole is a rotating disk of ionized gas that is oriented roughly perpendicular to the jet. This gas is moving at velocities of up to roughly 1,000 km/s. Gas is accreting onto the black hole at an estimated rate equal to the mass of the Sun every ten years.

Conclusion

Black holes retain their fascination despite the decades of solid research. They are both simple and complex: simple because it takes only three parameters to describe them; complex because it takes GTR to handle the dynamics. They come singly, in binary systems, and in pairs, but never 'naked'. They come both small and large in mass. They are impossible to see, but their effects on their environment can be distinctive, although they are not cosmic vacuum cleaners. We think they are found in the centers of most galaxies. There are dozens of possible detections of stellar mass black holes.

Gravity trumps all the other forces of nature in these objects. It compresses the mass of a dozen Suns, or a million, or a billion into a pinpoint of infinite density. Space and time are squeezed out of existence, and the structure of the universe turns into a "quantum foam" that's ruled by laws that scientists do not yet fully comprehend. We have a lot more to learn.

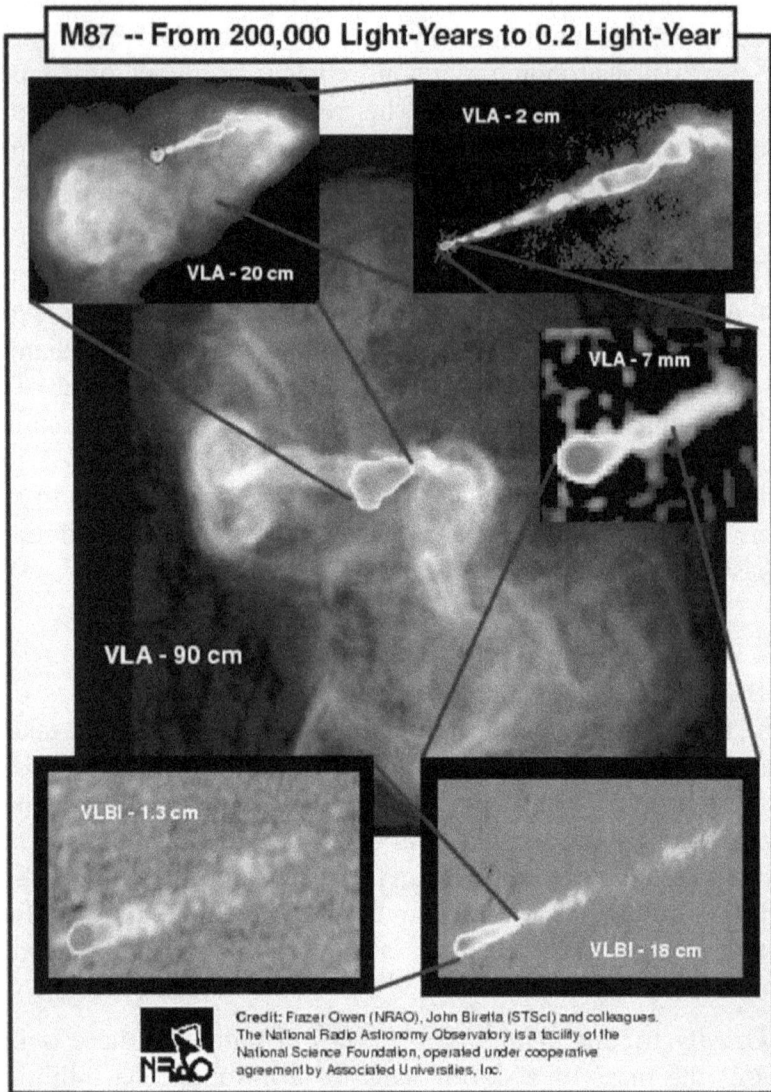

Figure 12. A series of multi-wavelength photos of M87 and its jets. Lobes of matter from the jet extend out to a distance of 250,000 light-years. Start in the center, then move to the upper left and follow clockwise the expansions of each image.

---

[i] The term 'black hole' was first publicly used in 1967 by physicist John Wheeler during a lecture. He always insisted that it was suggested to him by somebody else.

[ii] John Michell (1724 – 1793) was an English natural philosopher and geologist whose work spanned a wide range of subjects from astronomy to geology, optics, and gravitation.

[iii] Henry Cavendish FRS (1731 – 1810) was a British scientist noted for his discovery of hydrogen which he called inflammable air.

[iv] Michell, J. "On the Means of Discovering the Distance, Magnitude, &c. of the Fixed Stars, in Consequence of the Diminution of the Velocity of Their Light, in Case Such a Diminution Should be Found to Take Place in any of Them, and Such Other Data Should be Procured from Observations, as Would be Farther Necessary for That Purpose." *Phil. Trans. R. Soc. (London)* **74**: 35–57 (1784).

[v] Pierre-Simon, marquis de Laplace (23 March 1749 – 5 March 1827) was an astronomer/mathematician.

[vi] It represents the curvature in a Riemannian manifold. A tensor is a geometrical higher-order vector. Think of a matrix, although all matrices are not tensors.

[vii] Einstein called Λ his greatest blunder. Today scientists use it to explain 'dark energy'. It was originally introduced by Einstein to allow for a static universe (*i.e.*, one that is not expanding or contracting). This effort was unsuccessful for two reasons: the static universe described by this theory was unstable, and observations of distant galaxies by Hubble a decade later confirmed that our universe is, in fact, not static but expanding.

[viii] This allows one to measure intervals and to define distance in the curved space.

[ix] On the outbreak of war in August 1914 Schwarzschild volunteered for military service. While at the Russian front he wrote two papers on relativity theory providing the first exact solution to the field equations.

[x] A few months after Schwarzschild's work, mathematician Johannes Droste independently gave the same solution for the point mass.

[xi] Singularities are difficult to describe. They are absolute termination points – cessation of existence.

[xii] Actually Schwarzschild solved the equations with no mass and, then, in the weak field approximation, used the mass to bring it into coincidence with the Newtonian limit.

[xiii] Stationary means the black hole might rotate but not translate. A non-stationary black hole might be one that is orbiting another object.

[xiv] There are other metrics that are beyond the scope of this paper.

[xv] Bardeen J.M., Carter, B., Hawking, S., *Commun. Math. Phys.* **31**, 161-170 (1973)

[xvi] This means the infinity disappears in some coordinate systems.

[xvii] D. Finkelstein, Phys. Rev. **110**, 965–967 (1958).

[xviii] Which means we do not really know what happens.

[xix] It can orbit at this distance (and not fall in) if it moves quickly enough.

[xx] The $\sqrt{-1}$ is not allowed.

[xxi] *Science News*, **178**, p 28.

[xxii] See JWAS, Winter 2010 issue for a description of this experiment.

[xxiii] White holes cannot exist, since they violate the second law of thermodynamics.

[xxiv] Binary x-ray sources have a visible star and an invisible source of x-rays.

[xxv] Because angular momentum is conserved, observations of binary systems can give the total mass of the system.

[xxvi] 1 parsec is $3.08568025 \times 10^{13}$ km.